FROM DISCRETE TO CONTINUOUS

AUSTRALASIAN STUDIES IN HISTORY AND PHILOSOPHY OF SCIENCE

VOLUME 16

The titles published in this series are listed at the end of this volume.

KATHERINE NEAL
The University of Sydney,
Sydney, Australia

FROM DISCRETE TO CONTINUOUS

The Broadening of Number Concepts in Early Modern England

Springer-Science+Business Media, B.V.

Library of Congress Cataloging-in-Publication Data

ISBN 978-90-481-5993-2 ISBN 978-94-017-0077-1 (eBook)
DOI 10.1007/978-94-017-0077-1

Printed on acid-free paper

Originally published by Kluwer Academic Publishers in 2002.
Softcover reprint of the hardcover 1st edition 2002

For my grandmother Katherine Johnson and my daughter Elizabeth

TABLE OF CONTENTS

ACKNOWLEDGMENTS

I have received financial help for this book from the University of Toronto in the form of a May Fellowship and from the University of Sydney.

Several people have read this book in its many forms. I would like to thank Craig Fraser and Alexander Jones for their support in writing the original thesis, on which this book is based. The anonymous readers provided a great deal of valuable criticism, and I thank them for their detailed comments. I would also like to thank John Schuster for heroically reading the entire manuscript and John Henry for his encouragement and support. My thanks also go to Rachel Ankeny for making sure I actually had time to finish my revisions, and Jason Tong for his valiant efforts with the pictures, diagrams, formatting and index. Christopher Neal, my husband, has supplied invaluable emotional support. My daughter Elizabeth made my life happy but busy during the final process.

CHAPTER 1

TRANSFORMATION OF THE NUMBER CONCEPT

In the early modern period a crucial transformation occurred in the classical conception of number and magnitude.[1] For the Greeks, the unit, or one, was not a number; it was the beginning of number and it was used to measure a multitude. Numbers were merely collections of discrete units that measured some multiple. Magnitude, on the other hand, was usually described as being continuous, or being divisible into parts that are infinitely divisible. A distinction was made between arithmetic and geometry; arithmetic dealt with discrete or unextended quantity while geometry dealt with continuous or extended quantity.[2] In the early modern period a transformation occurred in this classical conception of number and magnitude. For instance, Simon Stevin (1548-1620) insisted in his 1585 *Arithmétique* that the traditional Greek notion of numbers was wrong: he believed that numbers were continuous rather than discrete.[3] Stevin also developed a system for indefinite decimal expansions of number that implicitly contained the idea of a numerical continuum.[4] Moreover, François Viète (1540-1603) introduced an improved form of symbolism:

---

[1] The term "number" is elusive and difficult to work with because its meaning changed often through time (this of course is one of the aspects that makes the subject historically interesting). Different meanings might be used by the protagonists of this book, secondary authorities cited such as Jacob Klein, and differently trained readers. For instance, mathematically trained readers may interpret 'number' as an element of the set of real numbers, defined either constructively (by the usual constructions including an equivalent of Dedekind cuts) from the natural numbers assumed given, or axiomatically by a set of axioms including some equivalent of a completeness property. These readers will therefore have a well-defined continuum in mind, in which the usual types of numbers as natural, integer, rational etc. are embedded. Readers without this mathematical background will probably have a different number concept in mind. Most likely this would be the common number concept of routine practitioners of letter-algebra: numbers are what letters stand for in algebra, they can be pictured along a scaled straight line (the so-called number line). Often the precise definition of these numbers is left to the experts, as is the precise definition of the global properties (such as continuity) of the number set. Less mathematically oriented readers usually locate the abstractness of numbers in the fact that it is not necessary to say what they are, and yet one can do algebra with them. In this book, terms such as numbers and continuity will be mainly used as defined by the actors in question.

[2] Olaf Pedersen, "Logistics and the Theory of Functions: An Essay in the History of Greek Mathematics," *Archives Internationales d'histoire des Sciences* 24 (1974): 29-50, has pointed out that logistics, or practical reckoning, although it was denied the status of being a theoretical science and thus not written about for its own sake, was carried to a greater state of perfection than has previously been suspected. For example, Ptolemy's *Almagest* depended upon numerical calculations whose methods and procedures were never explained in the text. However, no Greek practical mathematics texts were available in the early modern period.

[3] Continuous in an Aristotelian sense as will be discussed in chapter 2.

[4] Meaning in this case that Stevin noted that numbers can be pictured along a number line.

unknowns were distinguished from given magnitudes by being labelled with capital vowels while the given magnitudes were denoted by capital consonants. This powerful new symbolism was used to denote both unknown magnitudes and numbers; it indicated that numbers and magnitudes were, in a sense, interchangeable.[5] This association of numbers with magnitudes encouraged the notion that numbers could also be treated as though they were continuous in the Aristotelian sense. Thus the traditional idea of discrete number *versus* continuous magnitude was challenged in the early modern period in several ways. Stevin belonged to a tradition of computational mathematics that had a practical, problem-solving orientation. These practitioners were independent of the more theoretically oriented mathematics of the humanists.[6] The demands made by bookkeeping, accounting, measuring, navigation, and war had created dissatisfaction with traditional methods. Stevin began his career as a bookkeeper; his first work, published in 1582, was titled *Tables of Interest.* This work, which includes many examples of common business problems as well as tables of interest, shows that mercantile computations were becoming quite complex.[7] Stevin's 1585 edition of *The Disme,* dedicated to the wish that "any man's business may be performed easily thereby," was motivated in part by a desire to ease practical computations.[8] Although Stevin's development of an indefinite decimal expansion of number (with its implicit idea of a numerical continuum) did not *explicitly* become a part of the theoretical discussions about mathematical continua in Europe in this period, it certainly advanced the notion of numbers being continuous rather than discrete.

Viète, on the other hand, had a deep regard for the ancient Greek sources, although at the same time he was committed to algebra as a powerful problem-solving technique. However Viète's methods differed substantially from the earlier work of the abacus schools and of the cossic algebraists.[9] His emphasis was purely theoretical; he did not consider practical problems. Indeed, he went as far as giving up the name of algebra altogether.[10] Moreover, in the earlier works symbols only denoted numbers and they

[5] Later in the chapter there will be a discussion about the distinction between geometrical and abstract magnitude.

[6] For a wonderful discussion of the impact of humanism on mathematics in the 15th and 16th centuries, in particular the collection and translation of Greek mathematical authors such as Apollonius, Archimedes, Diophantus, Euclid, Hero, Pappus and Proclus, see Paul Lawerence Rose "Humanist Culture and Renaissance Mathematics: The Italian Libraries of the *Quattrocento*" *Studies in the Renaissance* 20 (1973)

[7] Simon Stevin, *The Principle Works of Simon Stevin*, ed. D.J. Struik, vol. 2 (Amsterdam: C.V. Swets and Zeitlinger, 1958). *The Principal Works of Simon Stevin* includes his *Tables of Interest.* In the early modern period banking houses pursued varied activities involving questions of insurance, of annuities and other payments at set intervals, of discounting of sums due at a later date and related transactions. Many of the examples in Stevin's book utilise fractions and complicated (for the period) interest computations. See also N. Z. Davis, "Sixteenth Century French Arithmetics and the Business Life," *Journal of History of Ideas* 21(1960): 18-48, and the discussion in Dijksterhuis' *Simon Stevin* for details.

[8] The dedication reads, "To Astronomers, Land-meters, Measureres of Tapestry, Gaugers, Stereometers in general, Money-Masters, and to all Merchants, SIMON STEVIN wishes health," *The Mathematical Works of Simon Stevin*, p. 389.

[9] Cossic algebra and Viète will be discussed in detail in chapter three.

[10] Viète believed that algebra, as it was practiced in his period, had been "spoiled and defiled by the barbarians." *Introduction to the Analytic Art*, trans. Reverend J. W. Smith, in J. Klein's *Greek*

played an essentially abbreviatory role. The earlier algebras had been devoted to finding 'recipes' or algorithms that would allow the solution of problems of various sorts, and no general formulas were derived. Viète's symbolism, on the other hand, allowed a more general treatment of problems and it focused attention on the procedures of the solution rather than on the solutions themselves. Thus there was pressure on the Greek conception of number and magnitude from both the practically oriented Stevin and the more theoretically focused Viète.

One difficulty in exploring such transformations is that in the early modern period mathematical terminology was still in the process of development[11]. It had recently been pointed out that terms such as "arithmetic", "algebra", "geometry", and "analysis" have modern meanings so different from their early modern counterparts as to make their use "downright misleading" without further explication. At this stage, it is important to settle such terminological issues, and I have attempted to remain as close to early modern usage as possible. Number, as already pointed out, was primarily construed in the classical sense of a multitude of units. Geometrical magnitude will refer to line segments, plane figures, and solid figures, when they were considered in terms of their size. My use of the term corresponds to the classical Greek conception of geometrical magnitude that was generally accepted in the sixteenth and seventeenth centuries. One must keep in mind that the size, say, of a line segment was not a number expressing the measure of its length with respect to some unit; size was its magnitude, which it shared with congruent line segments.[12] I will use the term *abstract magnitude* for "mathematical entities that, like geometrical magnitudes, could be joined, separated, and compared, but whose further nature was left unspecified."[13] Finally, ratio will refer to a relation with respect to size between two magnitudes of the same kind or between two numbers, not to magnitudes or numbers themselves. Another approach was available, however, in the early modern period that considered ratios in terms of "denominations". The denomination of a ratio of two numbers was depicted in terms of as a rational number in simplest form.[14]

Terms such as unknown, indeterminate, and variable are of a more recent date, and there are no early modern examples to draw upon. I will follow recent literature in using the term *unknown* to designate mathematical objects that, in an arithmetical or geometrical problem, were required to be determined by calculation or by construction. *Indeterminates* will designate mathematical objects whose particular size or value were

---

Mathematical Thought and the Origin of Algebra (New York: Dover Publications, 1992), pp. 318-319.

[11] I would like to thank an anonymous referee for introducing me to Henk J. M. Bos's magisterial *Redefining Geometrical Exactness: Descates' Transformation of the Early Modern Concept of Construction* (New York: Springer-Verlag, 2001). Much of what follows is taken from Bos's chapter six "Arithmetic, geometry, algebra, and analysis" pp. 119-132.

[12] Bos, *Redefining Geometrical Exactness*, p. 120.

[13] Bos, *Redefining Geometrical Exactness*, p. 121.

[14] See John E. Murdoch, "The Medieval Language of proportions: Elements of the Interaction with Greek Foundations and the Development of New Mathematical Techniques," in *Scientific Change* (ed) A. C. Crombie (London: Heinemann, 1963), pp. 237-271; E. D. Sylla "Componding Ratios: Bradwardine, Oresme and the First Edition of Newton's *Principia*," in (ed) E. Mendelsohn (Cambridge: Cambridge University Press, 1984), pp.11-43.

inessential to the final result.[15]

In the following, *arithmetic* will refer to the mathematical theory and practice that dealt with number. *Geometry* refers to the mathematical theory and practice that dealt with geometrical magnitudes. *Algebra* refers to those mathematical theories and practices that involved unknowns and/or indeterminates, employed algebraic operations, used equations, and dealt with numbers, geometrical or abstract magnitudes. *Analysis* is made up of mathematical methods for finding the solutions of problems or the proofs of theorems by introducing unknowns. Analysis may, or it may not, involve the use of algebra.[16]

Jacob Klein's seminal work, *Greek Mathematical Thought and the Origin of Algebra*, suggests an explanation for this transformation of the number concept in the work of such scholars as Stevin and Viète. He proposes that a conceptual transformation occurred that permitted a *symbolic* understanding of number, allowing numbers to refer to things other than numbered objects rather than always being associated with a "definite number of definite objects."[17] An increased interest in "practical, i.e. applied, mathematical disciplines," he implies, led to more elaborate techniques of calculation and the introduction of negative and irrational magnitudes. Additionally, he claims that the practitioners were unaware of the "novelty" of their number concept until they came into contact with the *Arithmetic* of Diophantus.[18] Moreover, he argues that treating numbers as symbols rather than as references to objects, was crucial to the emergence of modern mathematics.[19,20] It "obliterated" the classical distinction between continuous magnitude and discrete number.

It is not always altogether clear what Klein means by a more "symbolic" understanding of number. The difficulty stems from Klein's big picture orientation together with the wide range of mathematicians he discussed. I would argue that he supposed that number began to be seen as a kind of abstract magnitude, as something that could instantiate an algebraic symbol, especially an unknown that varies continuously. Unfortunately, Klein did not distinguish between abstract and geometrical

---

[15] Bos, *Redefining Geometrical Exactness*, pp. 127-128,

[16] These definitions have been taken from *Redefining Geometrical Exactness* pp. 128-130.

[17] Jacob Klein, *Greek Mathematical Thought and the Origin of Algebra* (Cambridge: M.D.T. Press, 1968), p. 7.

[18] Klein, p. 124., p. 148. Klein claims that the algebraic tradition, from Leonardo of Pisa *via* the "cossic" school, and up to such figures as Michael Stifel (1544), Cardano (1545), Tartaglia (1556-1560), was separate from the traditional disciplines of the schools, and it struggled for a place in the system of western science. This school, he further claims, only became aware of its own "scientific" character and the novelty of its "number" concept when it came into contact with the *Arithmetic* of Diophantus. He cites as an example Bombelli's modifying the "technical" character of his manuscript to a generalised form of the first five books of Diophantus after reading the *Arithmetic*. Moreover, practitioners such as Viète felt their work was an extension of a received ancient tradition because they believed that Diophantus had used a form of analysis similar to their own, but that Diophantus had hidden his method so that his "subtlety and skill might be more admired." It is unclear, however, in what sense researchers needed to be "self-conscious" in order to broaden their number concepts or how their self-awareness actually influenced the course of research.

[19] Klein, p. 125.

[20] Klein, p. 213.

magnitude, and it is not always apparent what definition of number is being utilised at any given point in the text.

This broadening of acceptable numerical results and the association of numbers with abstract magnitudes had a crucial impact on early modern mathematical development. It allowed numerical solutions to be offered for many problems in such areas as naval warfare, the expansion of trade, and merchants' accounts.[21] It also addressed the needs of customs officials, excise men, gaugers, and assayers. The increasing power of applied mathematics helped to develop public interest in mathematics; more students sought instruction in mathematics and more patrons were willing to support mathematicians. Several areas of theoretical mathematics such as the theory of equations, developments in the binomial theorem, and work with sequences and series, also benefited from the expanded conception of number. Moreover, this broadening of number concepts also led to attempts to supply new foundations for mathematical practice and new ways of perceiving the relationship between number and magnitude.

Scholars have explored this transformation for certain figures such as Stevin, Viète, and Descartes.[22] Klein, the foremost scholar in this field, partially attributes this change to the influence of practical mathematical disciplines. He does not, however, show how the actual problems being solved required a broadened number concept.[23] Few studies have gone beyond the mathematicians studied by Klein or traced the broadening of the number concept in the general mathematical practice of the early modern period.[24]

This book explores the blurring of the distinction between number and geometric magnitude and the development of new number concepts in early modern England and Scotland. No studies have been done specifically on this subject using English mathematicians. It will differ from prior research by examining this topic in a particular

[21] In some ways, of course, numerical solutions had always been available. As Pedersen pointed out in the context of Ptolemy, logistics was carried to a greater state of perfection than has previously been suspected. As we shall see, even scholars who maintained classical definitions of numbers often used fractions.

[22] See, for example, Jacob Klein's *Greek Mathematical Thought and the Origin of Algebra*; Charles Jones' "The Concept of One as a Number," diss. University of Toronto, 1978; Jones extends Klein's work by taking a more detailed look at how the classical concept of number and the Greek distinction between number and magnitude, was overthrown by Stevin. See also, Joann Morse's "The Reception of Diophantus' "Arithmetic" in the Renaissance," diss. Princeton University, 1981.

[23] For another interesting discussion, see: Michael Mahoney, "The Beginnings of Algebraic Thought in the Seventeenth Century," *Descartes Philosophy, Mathematics and Physics*, ed. Stephen Gaukroger (New Jersey: Barnes and Noble, 1980), p. 147. Michael Mahoney associates, like Klein, the broadening number concept in Viète and Descartes with the burgeoning algebraic mode of thought. Mahoney parts from Klein in attributing Viète's and Descartes' willingness to move beyond Greek mathematical concepts to the pedagogy of Peter Ramus and the search for a universal symbolism

[24] Helena Pycior, "Mathematics and Philosophy: Wallis, Hobbes, Barrow and Berkeley, *Journal for the History of Ideas* 42 (1987): 263-286, does briefly mention the unusualness of Isaac Barrow's number concept, and she attributes Barrow's point of view to his recognising only geometry as a mathematical science. The main focus of her article, however, is the seventeenth-century roots of Berkeley's mathematical views. Thus she does not explore the details of how the dispute between ancients and moderns influenced the transition to an algebraic mode of thought. C. Sasaki, "The Acceptance of the Theory of Proportion in the Sixteenth and Seventeenth Centuries," *Historia Scientiarum* 29 (1985): 83-116, notes the differences between Wallis' and Barrow's views on algebra and number, and he relates their differences to a propensity towards 'modern' *versus* 'traditional' points of view.

mathematical community for a specific period, from the introduction of algebraic techniques to the first attempts to formulate justifications for the broadened number concepts and their changing relationship with magnitude. Moreover, this book will also further the investigation of this topic by focusing on mathematical practice and how the desire to solve particular problems helped to motivate the acceptance of the broadened number concept. For example, the adoption and development of certain mathematical methods, such as logarithms, carried with it certain tacit theoretical implications about the nature of numbers. Klein's explanation, however, will also be utilised by exploring the link between the use of algebraic symbolism and a more "symbolic" understanding of number in English mathematics. As algebra began to be used to treat number, figure, and mechanical entities it brought into being this notion of abstract magnitude. Abstract magnitude can then be considered a to be a genus that includes as species generalised number, geometric magnitude, and (for example) force and speed.[25] Klein's assumption of the importance of contact with Diophantus' *Arithmetic*, on the other hand, does not appear to have been accurate with respect to the English and Scottish mathematical communities.

These scholars were aware of the classical theoretical issues and some of their work was influenced by the advances made by Viète, although they worked mainly in the practical tradition of Stevin and the earlier abacus algebras. The change in the English and Scottish notions of number and the shift from geometrical magnitude to abstract magnitude came partially from the demands of practice. The types of problems being solved, however, changed over time. They ranged from practical activities such as measuring and commerce, to the expression of a quadratic equation that captured the essence of a Greek geometrical problem, to topics in free fall and optics. Some developments, such as the logarithms, were associated with a practical concern with navigational techniques. It is only with the work of John Wallis and Isaac Barrow that we see a direct interest in theoretical concerns relating to a reworking of classical geometry problems and more advanced numerical methods.

Mathematics was not a unified discipline in this period. There were several subdisciplines, each following its own characteristic patterns. At least three different approaches existed. First were the algebraists, whose work centred on the new algebraic methods and the theory of equations. Robert Recorde (1510-1558), William Oughtred (1575-1660), and Thomas Harriot (1560-1621) all produced work in this tradition. A second group was made up of such figures as John Napier (1550-1617) and Henry Briggs (1561-1630), who were influenced by an interest in practical computation techniques, and who developed the logarithms. Their construction of logarithms treated numbers as being interchangeable with points on line segments. Finally, there were the university mathematicians, Isaac Barrow (1630- 1677) and John Wallis (1616-1703), who while attempting to work within the ancient tradition, were at the same time inspired by more modern developments. Barrow's work provides us with a glimpse at how numbers and magnitudes were viewed by a scholar who was mainly concerned with geometry. Wallis, on the other hand, was interested in the new analytic methods and

[25] This is an observation made by an anonymous referee.

infinitesimals. The heterogeneous approaches to abstract magnitude by these subdisciplines compel a topical rather than a chronological organisation. Examining all three groups will allow us to explore a broad range of English and Scottish mathematical practice.

A study of these developments in England will allow us to glimpse the process of broadening of number concepts and the breakdown of the sharp distinction between number and geometric magnitude and the shift towards a notion of abstract magnitude in a fairly isolated mathematical community. By beginning with Recorde, we can investigate what the views of number and magnitude were like before the English learned about the work of Viète and Stevin. Moreover, with Recorde, the English begin slowly to assimilate the continental cossic symbolism. Continental mathematics texts, however, were sometimes difficult to obtain in England. Although there was an awareness of certain continental works, there were also gaps and barriers in the mathematical interaction between Britain and Europe until Wallis' and Barrow's time, which marks the end of our study. For instance, although the logarithmic tables were quickly assimilated by Europeans, there was little interest in Napier's methods, and although figures such as Oughtred were interested in continental developments, they were often unable to obtain European texts. Wallis and Barrow, on the other hand, were professional mathematicians with a wide-ranging knowledge of mathematical developments. With their work, we can explore the first attempts to justify the broadening number concepts and the new ideas about the relationship between number and magnitude. Their work also shows a merging of the English and Continental traditions. Although they were influenced by the internal English developments, a wider range of mathematical practice also affected them. Finally, a study of English and Scottish practice will allow us to investigate what factors most influenced the broadening of number concepts and the transformation of the relationship between number and magnitude in a reasonably homogeneous culture.

In early modern England and Scotland mathematics was being called upon, as it had been called upon earlier in Europe, to answer questions that it had not been asked before. Tudor England, in particular, was economically and organisationally backward in comparison with the best practices of continental Europe. It was not until the middle of the sixteenth century that England began to reach western European standards of development. For instance, it was not until 1553 that corporations such as the Muscovy company formed, creating a greater demand for the abacus-style merchant mathematics that had been adopted in Italy as early as the fourteenth century. In this period, there was also an increased interest in England and Scotland in practical mathematical applications such as ballistics, navigation, and surveying. The solutions for some of these problems, as we shall see, called for the use of fractions and irrational numbers. Thus we will examine how the particular demands of local practice influenced this topic.

We will also investigate Klein's thesis that the adoption of algebraic methods is connected to a more symbolic view of numbers and a breakdown in the traditional separation between number and magnitude. Early modern England saw the beginning of a movement from an abbreviatory type of symbolism, in which each power of a variable required a separate symbol, to a completely general symbolism, in the style of

Viète and Descartes. This improved notation helped create the idea of an abstract magnitude. Unknowns came to be seen as symbols that could stand for numbers, geometric magnitude, or mechanical entities, and this fact weakened the distinction made by the Greeks between discrete number and continuous geometric magnitude.

The book will also investigate how the broadening of number concepts and the transformed understanding of the relationship between numbers and abstract magnitude influenced English mathematical practice. Europeans had earlier developed a practical, numerically based, mathematics that was separate from, but informed by, the theoretical mathematics that emerged in the wake of humanism. In this period the English and Scottish also formulated practical numerical methods that were not theoretical themselves, but that showed an awareness of theoretical issues. Even the earliest, and relatively unsophisticated, mathematical practice of Robert Recorde displays a tension between the traditional view of number and geometrical magnitude and the need for a wider variety of numerical solutions. Although Recorde stated that "Unitie is of it self indivisible," and that unity is the origin of number, and followed Aristotle by having the printers place in the margin of the text that "One is no nomber," he was able to utilise several types of numbers. For instance, he was willing to work with the positive rationals, which he called broken numbers, and irrationals, which he called surds, although he was uncertain about their status as numbers. Other scholars, such as Harriot, displayed their interest in these problems by questioning Greek ideas; his papers contain a work titled *De infinitis* that directly queried Aristotelian ideas about continua. He considered several questions: is the infinite composed of finite parts? Further, is the finite composed of indivisible parts and can it be resolved into indivisibles? Is there a transition from the finite to the infinite through a maximum finite? Harriot's unpublished manuscripts show that some British scholars went beyond a concern with a more general number concept to directly consider the nature of continua.

More often mathematicians showed an interest in broadening the nature of numbers without explicitly discussing abstract magnitude. Napier, for example, wrote an unpublished manuscript on algebra that displays a great concern with number concepts. He was willing to acknowledge negative numbers, fractions, and irrationals. Moreover, he was able to justify this acceptance by stating that it was only pure mathematics that was limited to discrete numbers. Additionally, he constructed distinctions between types of numbers. For example, he stated that "geometry is the logistics of material quantities by the use of concrete numbers. . . . A number is called a concrete number insofar as it relates to a material and continuous entity." All irrational numbers were labelled concrete because no fractional numbers are commensurable with them. Most of the discussions that considered the nature of numbers and the relationship between number and magnitude were embedded in texts dealing with various aspects of mathematical practice. They also show an interest in broader theoretical issues such as the nature of number and its proper relationship with magnitude.

The highlight of this exploration of the impact of algebra is the discovery that an interest in solving practical problems was not enough to cause a broadening of the number concept. It was combined impact of both novel practical applications and algebraic notation that produced this mathematical innovation. The examination of logarithms shows that as the purpose and goal of mathematical research for practitioners such as Napier and Briggs lay in its practical applications, they were more than willing to violate the canons of Greek mathematical philosophy

After we have explored how these concepts were treated in English and Scottish mathematical practice, the book will conclude with Isaac Barrow's and John Wallis' extended discussion of the nature of number and geometric magnitude. We will also be able to consider Barrow's view on the composition of magnitude, space, and time. This will allow us to investigate the first English attempts to supply some kind of justification for the broadening of number concepts and also give us a glimpse at various ways that number was related to magnitude. Moreover, since Barrow and Wallis had widely different interests and used distinct mathematical techniques, we will be able to see how their separate conceptions of these mathematical entities related to their mathematical practice.

The first chapter will begin with an overview of the traditional Greek ideas about numbers and mathematical continua such as magnitude, space, and time that were available to early modern English mathematicians. The concepts of Aristotle, Euclid, Proclus, and the Greek atomists will be reviewed in order to provide a framework for our study. This overview will also allow us to discover when the mathematicians in this study were presenting original views by departing from the traditional ideas. The second chapter will explore the relevant contemporary mathematical methods and opinions that were available to the practitioners in our study. We will consider how the changing nature of the problems being solved and the development of new symbolism and techniques influenced English developments. In particular, we will explore what aspects of algebraic practice might have influenced English ideas concerning number and its relation to magnitude. We will investigate what developments helped these scholars to broaden their number concepts and move towards an appreciation of abstract magnitude, and how they helped them to reject the difference in *kind* between numbers and magnitude postulated by the Greeks.

The third chapter will examine the English algebraists. Since mathematicians did not often discuss their conception of number and geometric magnitude we will use their works to examine how algebra, as a new mathematical tool, helped to shape the way in which numbers were depicted and how it influenced views on the possibility of utilising a notion of abstract magnitude. Several early modern English algebraic texts will be explored in our investigation of how algebraic techniques and problem choices might have shaped numeric representations and influenced views as to the relation between numbers and magnitude. First, we will study Robert Recorde's 1557 work, *The Whetstone of Witte.* This text is the earliest English algebra and it contains practical problems whose answers could not be expressed as positive integers. Moreover, the *Whetstone* also gives us an idea of the level of algebraic knowledge first introduced into England. Some of the other algebras to be examined will be William Oughtred's *Clavis Mathematicae* (1631) and Thomas Harriot's *Artis Analyticae Praxis* (1631). These texts provide us with an example of English algebra at a more developed stage.

The fourth chapter will investigate the "practical" mathematicians who created the logarithms. The logarithms, invented by John Napier and developed by Henry Briggs, were important to the evolution of ideas about the relationship between number and magnitude and the broadening of number concepts for several reasons. Logarithms were always depicted in terms of decimal expansions, and this implies that their creators did not view numbers as being collections of discrete units. Moreover, Napier's conception of the logarithms explicitly maps numbers to points on a line segment. Although he did not utilise the variables introduced by the algebraists, he seems to have considered numbers and geometric magnitudes as being interchangeable. The logarithms will also show how practical concerns such as an interest in improving navigation could lead to the acceptance of new kinds of numbers.

The chapter focusing on logarithms will discuss Napier's (1616) *Description of the Admirable Canon of Logarithms* and its 1616 translation by Edward Wright. Moreover, we will explore Napier's *Mirifici Logarithmorum Canonis Constructio*, which was first published in 1619, two years after his death. This text provides Napier's original conception of the logarithms and shows how he used kinematic techniques to link magnitudes to numbers. We will also examine the 1631 posthumous English translation of Briggs' work titled, *Logarithmicall Arithmetike*. This book is not an exact translation of Briggs' earlier work, but it is of interest because it leaves out theoretical details in order to give a wide variety of practical examples that show the usefulness of logarithms to such fields as the "true Measuring of Lands, Plots, Buildings and fortifications, Astronomie and Dialling, Geography and Navigation."

The final chapters of this book will explore the views of Isaac Barrow and John Wallis on the relationship between numbers and magnitude and mathematical continua. Barrow and Wallis each had their own particular mathematical methods, and they also had different interests, so they will be discussed in two separate chapters. Through the investigation of Barrow and Wallis we become familiar with the points of view of professional mathematicians who favoured "geometric" and "algebraic" practice respectively.

Barrow was one of the few mathematicians who discussed his conception of geometrical magnitude and the relationship between numbers and magnitudes in any detail. We will explore the nature of his mathematical practice though an investigation of his translation of Book II of Euclid's *Elements*, which was often considered "algebraic," and through his *Geometrical Lectures*. His translation of Euclid's *Elements* (1655) was his first mathematical publication, and his version of Book II will provide us with his earliest views about the relationship between variables, numbers, and magnitudes. His *Geometrical Lectures*, first published in 1670, will supply us with an idea of Barrow's mature mathematical style. Knowledge of Barrow's mathematical preferences and methods will help us to understand his conception of the relationship between magnitude and number and the nature of mathematical continua. Barrow's work, *The Usefulness of Mathematical Learning,* was derived from a three-term series of lectures begun when he was awarded the Lucasian Professorship. In this text he developed a new foundation for numbers and gave them a nontraditional relationship to magnitude. This text will also allow us to investigate Barrow's views on the nature of space and time. Because of his interest in curves generated by motion, he explored these topics in unusual detail.

An examination of the mathematical practice and views on the relationship between numbers and magnitude of John Wallis will show that there were several options available for justifying the broadened concept of number and the breakdown in the separation between number and magnitude. Although he did not explicate his views on continua concepts in as much detail as Barrow, his point of view is of interest as it was closer to those of the algebraists we examine earlier in our study. Wallis' mathematical practice will first be explored through an examination of some of his translations of Book II of Euclid's *Elements*. We will also briefly investigate Wallis' *de sectionibus conics,* in order to examine how Cartesian methods and the method of indivisibles influenced him. Finally, we will explore his best known work, the *Arithmetica Infinitorum*, to illustrate Wallis' facility for finding numerical patterns. This section will also discuss his "method of induction" and "principle of continuity." A picture of Wallis' mathematical research will help supply us with his motivation for his position on the nature of numbers and mathematical continua.

Finally, we will investigate the various places where Wallis discussed the relationship between numbers and abstract magnitude and his concept of geometrical magnitude. For example, sections of his *Mathesis universalis* were devoted to explanations of number concepts. Wallis' views on the nature of numbers, however, were not presented in one place in a systematic fashion. Indeed, sometimes he expresses conflicting opinions. His ideas, however, will provide us with the point of view of a professional mathematician whose practice was mainly algebraic.

Overall, this book will show how numbers were used in a variety of different subdisciplines. Since the classical point of view was seldom directly addressed in any detail, actual mathematical practice is our best guide to the changing concepts of number and the new ideas about their relationship to magnitude. The scholars in this study often officially deferred to the classical dictates on the nature of numbers. Nevertheless, they were sometimes forced to expand the ancient number concept in order to solve a particular problem. A tension existed between their awareness of classical theoretical concerns and the requirements of problem-solving. We will explore two factors in particular that influenced this gradual acceptance of fractions and irrational numbers as well as the breakdown in the traditional distinction between number and magnitude. First, there was the changing nature of the problems being solved and the development of new mathematical methods that encouraged the broadening of acceptable solutions. Second, the increased use of an improved variable notation inspired the notion that numbers and magnitudes were interchangeable. At the end of our study we will see that although no consensus was reached in this period about the numerical status of fractions and irrationals or the relationship between numbers and magnitude, these numbers were fully integrated into mathematical practice: they were just too useful to ignore.

CHAPTER 2

THE ANCIENT SOURCES

This chapter will provide an overview of the ideas concerning mathematical continua and related topics such as the nature of number that were available to early modern English mathematical practitioners. This summary is intended to furnish a yardstick by which we can measure the departures of early modern mathematicians from standard classical opinions. Moreover, it will make it unnecessary to recapitulate, for example, Aristotle's views on continua every time a comparison is made between his ideas and an early modern concept.

First, classical Greek notions of continua and numbers will be investigated. Not all classical views will be addressed, only those ideas that might have been familiar to the early modern British mathematicians will be explored. Thus the section describing classical notions is not meant to provide a thorough outline of all the various opinions formulated by the Greeks, since much of this material was unavailable to the scholars in this study. It will be shown that Aristotle's opinions were fairly common knowledge. In addition, Euclid's *Elements* was familiar to all of the mathematicians in this study, while Proclus' views were at least available to those who sought them out. Atomistic ideas seems to have only influenced Thomas Harriot, out of the practitioners in this study, although they were at least discussed by Isaac Barrow, and they will be the last ancient opinions discussed in this chapter.

ARISTOTLE

Many of Aristotle's works, especially his books on natural philosophy, were an important part of the curriculum of early modern England's universities. The English philosopher John Case (ca. 1540 – 1600), who was active at Oxford, wrote popular expositions on Aristotle's logic, physics, and moral philosophy that were frequently used texts.[1] Johannes Magirus' *Physiologia peripatetica*, first published in 1597 and reprinted seventeen times by 1642, was another work widely used in the seventeenth century at Cambridge and Isaac Newton numbered among its readers.[2] But it is difficult to trace the actual texts used by all the mathematicians in this study. The number of translations of Aristotle's works was enormous and there were also numerous detailed philosophical commentaries that drew on wide ranging material relating to Aristotle, everything from Philoponus to the Merton school.[3] Moreover, in this period medieval

---

[1] Charles B. Schmitt, *Aristotle and the Renaissance* (Cambridge: Harvard University Press, 1983), p.138.

[2] Schmitt, p. 52.

[3] Johannes Philoponos, from the first half of the sixth century A.D. specialised in original commentary on

commentaries were still being published. Thus although we know that almost all the mathematicians whose work we will examine were university men, and that they must have had varying degrees of familiarity with Aristotle's ideas about numbers and geometrical magnitude, we do not know in all cases how these opinions were conveyed

Robert Recorde, for example, was educated at Oxford, where he completed his B.A. and perhaps his M.A., and he was elected a fellow of All Souls College in 1531. He received the degree of Doctor of Medicine from Cambridge in 1545. Aristotle is often cited in Recorde's various works, although Recorde is aware of the danger resulting in an unreasonable adherence to authority, stating:

> It is commonly seene that when men will receive things from elder writers, and will not examine the thing, they seeme rather willing to erre with their auncients for company, than to be bold to examine their workes or writings. Which scrupulositie hath ingendred infinit errors in all kinds of knowledge, and in all civill administration, and in every kinde of art.[4]

Recorde, however, seemed to share certain of Aristotle's mathematical ideas as he had the printer type in the margin of *The Whetstone of Witte* "One is no nomber." Perhaps he was more willing to follow ancient authority in mathematics where there could be no empirical evidence either way.

While Recorde both cites Aristotle and seems to adhere to portions of his mathematical philosophy, William Oughtred, on the other hand, does not mention Aristotle in his books. Yet Oughtred attended King's College, Cambridge, receiving both a Bachelor and a Master of Arts, and it is therefore unlikely that he had not studied Aristotle's *Physics*. Thomas Harriot wrote an unpublished treatise that addressed Aristotle's discussion of continua in the *Physics*. Overall, Harriot showed a great familiarity with Aristotle's ideas, although he disagreed with many of them. In conclusion, the three writers of early modern algebras who will be discussed in this thesis all were reasonably well acquainted with Aristotle's work.

When we turn to the developers of the logarithms, the situation changes somewhat. John Napier never attended an English University and the schooling he received in Scotland seems to have been at a less than advanced level. He did, however, study in Europe and he might have examined Aristotle's writings there. Yet his work shows no signs of being influenced by Aristotle. Napier utilised decimal expansions and did not treat numbers as collections of units. Moreover, he also used what we would label "instantaneous velocity" in his model and Aristotle did not believe that motion occurred in an instant. Henry Briggs, on the other hand, although he also used decimal expansions, was a Cambridge graduate who was probably familiar with Aristotle's work. Neither Napier nor Briggs seemed overly concerned about abandoning the idea

---

Aristotle. His work provides an early exposition of impetus theory. See, Olaf Pedersen, *Early Physics and Astronomy: A Historical Introduction* (Cambridge: Cambridge University Press, 1993), p.359. Merton College, Oxford was famous in the 14th century for its 'calculators'. See Pederson, pp. 191-201.

[4] Robert Recorde, *Grounde of Artes*, p. 342 as quoted in F.R. Johnson's and S.V. Larkey's "Robert Recorde's Mathematical Teaching and the Anti-Aristotelian Movement," *The Huntington Library Bulletin* 7 (1975), pp. 59-87.

of numbers being collections of discrete units.

In the mathematics of Isaac Barrow and John Wallis, however, we see that the traditional Greek notions of number and geometrical magnitude influenced them both. Barrow's tutor, James Duport, was an Aristotelian who opposed the faction at Cambridge that tended to worship Ramus and Descartes. Duport's manual for students commended Aristotle for specific purposes.[5] Indeed, at one point in Barrow's career he was the Regius Professor of Greek at Cambridge, and he not only studied Aristotle's works, but he taught them as well. In his work, *The Usefulness of Mathematical Learning,* Barrow showed his knowledge of Aristotle's dictates on the nature of geometrical magnitude. Although Barrow did not always agree with Aristotle, he certainly studied his works and valued his opinions. Wallis, although he did not cite Aristotle directly, seemed to be influenced by traditional notions of numbers and mathematical continua. Since he earned both a B.A. and an M.A. at Cambridge, he was likely exposed at some point to Aristotelian ideas. As we shall see, in some ways Wallis seemed more concerned than Napier and Briggs to remain within the framework of classical ideas, at least in regard to his expression of number.

Of course, since Euclid used the same definition of number and unit as Aristotle's, some of these scholars might have been primarily influenced by Euclid. All of the mathematicians mentioned above were very familiar with Euclid's *Elements,* and thus we must next examine what portions of Euclid might have influenced their ideas on mathematical continua. Euclid did not give a definition of magnitude nor did he directly discuss the nature of mathematical continua, but he appears to have used the strict division between number and magnitude created by Aristotle to separate the geometrical material of his first six books from the arithmetic material of Books VII through IX.[6] Since the treatment of number and magnitude in Euclid seems to mirror to a certain extent that of Aristotle, a brief summary of Aristotle's ideas on these topics will be provided first, and secondly an examination of Euclid's opinions. Another reason for beginning with Aristotle's work is that he often described the Atomists' ideas in the process of refuting them. Therefore one way that a modern who was influenced by atomistic ideas, such as Harriot, could have learned about Atomistic ideas would have been through reading Aristotle.

The distinction Aristotle drew concerning arithmetic and geometry was that arithmetic dealt with discrete or unextended quantity while geometry dealt with continuous or extended quantity. The methods from one subject could not always be used on the other; they were to be considered using separate techniques. Aristotle stated "You cannot therefore when proving a thing pass from one genus to another; e.g. you cannot prove a geometrical proposition by arithmetic."[7] The discrete quantity operated on in arithmetic was number. For Aristotle the unit, or one, was considered to be a

---

[5] Mordechai Feingold, "Isaac Barrow: Divine, Scholar, Mathematician" in *Before Newton: The Life and Times of Isaac Barrow*, ed. Mordechai Feingold (Cambridge: Cambridge University Press, 1990), p.13.

[6] Charles Jones, "The Concept of One as a Number", diss. University of Toronto, 1978, p.57.

[7] *An. Post.* I. 6-7. 75^a35^b17, quoted in Sir Thomas Heath's *Mathematics in Aristotle* (Oxford: Clarendon Press, 1949), p.44.

substance without position; it was the beginning of number and it was used to measure a multitude, but it was not itself a number. He stated, "The least number, without qualification is the two. . . ."[8] Numbers themselves were merely collections of units that measured some multiple.[9]

Aristotle also explored the nature of numbers in his examination of the infinite. He stated, "in number there is a limit at the minimum, but in the direction of 'more' [number] always exceeds any multitude...number is a plurality of ones, a certain 'many' of them. So there must be a halt at the indivisible...But in the direction of more it is always possible to conceive of [more]."[10] Thus no matter how large a number one thought of, it was always possible to think of an even larger number by adding a unit. However for Aristotle there could be no "actual" infinite; these numbers could not actually all exist at the same time, but they could exist "potentially."[11] Aristotle's concept of continuity, or geometrical magnitude, is also developed in the context of his discussion of infinity.

Change was thought to be continuous, and Aristotle thought that a study of the infinite was vital to understanding both change and continuity. There could be no actually infinite body because of his theory of "natural place". There could, potentially at least, be an infinite with respect to division; he stated that "magnitude is not in actual operation infinite; but it *is* infinite in division--it is not hard to refute indivisible lines--so that it remains for the infinite to be potentially."[12] The two types of potential infinities, by adding or division, were mirror images of each other; and if one counted each division made then one would have the infinite by addition. The concept of potential divisibility was crucial for one of Aristotle's definitions of geometric magnitude; he stated in the *Physics* Book VI that "[A magnitude] is divisible into divisibles that are infinitely divisible." In other words, no matter how many times one divided a magnitude one never reached an indivisible; division could always (at least theoretically) be continued.

Aristotle defined 'continuous' more rigorously in Book V of the *Physics*. He began by defining *consecutive*; *A* is consecutive to *B* when it comes after it in position or some other kind of way and there is nothing between *A* and *B* of the same kind. Next, *A* is in *contact* with *B* when the 'limits' of the two things are in exactly in the same place. So, *A* is *continuous* with *B* when they are next to one another and in addition the limits at which they touch are the same thing, or "having become one", are being "held together." One way of looking at this would be that things that are merely next to each other do not necessarily hold together during motion, but if two things are continuous they move together. Aristotle claimed that it follows from this definition that a

---

[8] *Physics* 12 220^a 27.
[9] *Metaph. N. 1* 1088^a 4-8.
[10] *Physics* Book 3 207^b1.
[11] William Charlton, "Aristotle's Potential Infinities," in *Aristotle's Physics: A Collection of Essays*, ed. by Lindsay Johnson (Oxford: Clarendon Press, 1991), p. 129.
[12] *Physics*, 6 206^a 9 16-18.

continuum (such as a geometrical magnitude) cannot be made up from indivisible things (such as points).[13]

Many reasons were given to show that a continuum cannot be composed of indivisibles. First, an indivisible has *no* extremities or limits, because a limit must be different from what it limits. So what has a limit must have parts, and must therefore be divisible. Secondly, points, for example, cannot compose a line by being in contact because the whole of one point would have to touch the whole of the other, and since they have no parts and this would imply that the points were in the same place, but what is continuous must have parts in separate places. Moreover, a point cannot be consecutive to a point nor a moment to a moment because there is a line between any two points and a time between any two moments.

Aristotle also wanted to demonstrate that time and space are both infinitely divisible, and hence continua. He stated that "time is not thought to be composed of nows. . . for we take it as impossible for the nows to be adjoining one another, as it is for a point to be adjoining a point."[14] These "nows" can be viewed as points, or perhaps time atoms. Time implied change. The primary continuum is spatial magnitude, but movement is continuous as it is movement through continuous space, and therefore time is continuous since continuous movement occupies it. Aristotle stated,

> Now, anything that is going on has a start and comes to a stop, and any magnitude is continuous; and so a "movement" corresponds to a magnitude. Since a magnitude is continuous, so is a movement; and because of the continuity of the movement, time is also continuous, for we estimate how much time has passed by the amount of movement that has occurred. [15]

A lapse of time is for Aristotle only recognisable because of some change or motion. He decided that time was not change itself, but the way in which change has a number. But time is not number in the sense of "pure number", it is "that which is counted and not that by which we count. . . . Time is the number of motion, and the now is, as the moving thing is, like a unit of number."[16] The last idea that Aristotle had about time that might have influenced early modern English and Scottish mathematicians is that one was not permitted to say that a thing was changing at an instant, indeed, there was to be no rest, movement, or any other change in an instant. Thus if one adhered to Aristotle's ideas there could be no notion of instantaneous velocity.

Two main Aristotelian ideas might have influenced early modern mathematicians-- the discreteness of number and the nature of geometric magnitude. Only Isaac Barrow seemed interested in Aristotle's reasoning about the continuity of time. If they accepted the Aristotelian dichotomy between discrete number and geometric magnitude, neither fractions nor irrationals would have been considered numbers. The advent of new mathematical methods such as algebra and logarithms, however, made it difficult to

---

[13] For a very detailed discussion of Aristotle's views, see David Bostock's "Aristotle on Continuity in *Physics* VI," in *Aristotle's Physics: A Collection of Essays*, ed. Lindsay Johnson (Oxford: Clarendon Press, 1991), pp. 180-212.

[14] *Physics*, 6 218^a 7 and 218^a 15.

[15] *Physics*, 6 219^{a}10. Also see Sir David Ross's *Aristotle* (London: University Paperbacks, 1964), pp.89-90 for a clear discussion of Aristotle's notion of time.

[16] *Physics*, 6 219^{b}1 and 220^{a}3.

avoid solutions from the entire realm of real numbers. Moreover, exact solutions for some of the practical problems they wished to solve from fields such as navigation and surveying also took the form of real numbers. Most mathematicians, even those like Recorde who had stated that one was not a number, actually used a variety of numbers and merely gave them more specific labels such as "broken" or "surd" (irrational) numbers. Barrow, for example, accepted the real numbers and avoided any difficulties by describing them as mere "notes" or "signs" of magnitude. In a few cases, however, there was no discussion of the entire problem. They simply used whatever numbers were necessary without any contemplation.

The second form of possible Aristotelian influence was his definition of geometric magnitude and continuity. Acceptance of his ideas might have led to hesitation about using methods employing indivisibles, as no magnitude could be anything but infinitely divisible and could not be composed of indivisibles. Moreover, Aristotle provided a framework for discussion of continuity. Those few mathematicians, such as Harriot, who were influenced by atomistic views, felt it necessary to at least consider Aristotle's objections. However some of what was known about atomist ideas came from Aristotle's refutations of their ideas. For example, Aristotle provided Democritus' argument for indivisible magnitude in the *Physics*. Democritus had tried to establish this thesis by proving that its contrary (that magnitude is divisible everywhere) was false, as follows. If it was divisible everywhere, what would be the end products of such a division? It cannot be parts with magnitude, because such parts would be further divisible. Nor could it be parts without magnitude, because such parts could not be added up to make a whole magnitude. Democritus could see no third way so he concluded that there must exist some indivisible magnitudes.[17] Mathematicians could either agree with Aristotle's refutations of atomist positions or find arguments that were more compelling than the arguments of the original atomists.

EUCLID

There is no universal agreement as to whether Euclid was influenced by Aristotle or by previous geometry.[18] Indeed, earlier geometrical works might similarly have inspired both scholars. Nevertheless, there seem to be traces of the Aristotelian distinction between discrete number and continuous magnitude in the *Elements*.[19] Since all of the mathematicians in this study were familiar with Euclid's *Elements,* it is necessary to

---

[17] David Furley,"Aristotle and the Atomists on Infinity," in *Cosmic Problems: Essays on Greek and Roman Philosophy of Nature* (Cambridge: Cambridge University Press, 1989), p.104.

[18] Richard McKirahan Jr, *Principles and Proofs: Aristotle's Theory of Demonstrative Science* (Princeton: Princeton University Press, 1992), p.135.

[19] Charles Jones, in his thesis, "The Concept of One as a Number" (full citation above) makes an interesting case that Euclid was working in the tradition of Aristotle to develop a theory of incommensurability. He believes that Euclid's mathematics was complementary to Aristotle's philosophical analysis. Moreover, Jones discusses in greater detail the same Aristotelian and Euclidean material summarised in this section.

briefly summarise the relevant portions of the work in order to discover the influence of Euclid's notions. When we turn to the mathematical works of Barrow and Wallis, it will be shown that Euclid's work was viewed differently depending upon the interests of the mathematician. This section, however, will provide a framework for examining these later interpretations of the *Elements*.

Euclid provided his definition of unit and number in Book VII. Indeed the *Elements* mainly dealt with numbers in the arithmetic Books VII-IX, and in Book X where Euclid showed that the old proportion theory based on numbers could be applied to numbers only if they are commensurable. The definition he gave for **unit** was a "**unit** is that by virtue of which each of the things that exist is called one", while a **number** "is a multitude composed of units."[20] Thus Euclid, in his definition of number, agreed with Aristotle, ruling out viewing the unit as a number itself. The definition of unit does not seem to have been supposed to forbid its division, since the third and fourth definitions leave open the possibility that Euclid accepted proper fractions by stating "A number is **a part** of a number, the less of the greater, when it measures the greater" and "but **parts** when it does not measure it."[21] Heath claims that Euclid meant by **parts** any number of submultiples making up a fraction less than unity. Proposition VII, 4 also seems to indicate that Euclid condoned the use of fractions; it stated "Any number is either a part or parts of any number, the less of the greater." A step in the proof of this proposition seems to use the idea of proper fraction. Therefore while Euclid appears to agree with Aristotle about the nature of units and the definition of number, the *Elements* differs from Aristotle in its seeming acceptance of proper fractions, although Euclid never directly stated that proper fractions were numbers.

Book VII also includes a separate proportion theory for numbers that was not subsumed under the more general theory of Book V. Definition 20 stated, "Numbers are **proportional** when the first is the same multiple, or the same part, or the same parts, of the second that the third is of the fourth."[22] Euclid does not, however, provide a separate definition of the *ratio* of numbers, yet the term arises in the propositions without any explicit definition.[23] Some commentators who suppose that number was a particular case of magnitude are puzzled by Euclid's presentation of a separate proportion theory for numbers. But as Stillman Drake so aptly said, "Classification of what is not indefinitely divisible as a particular case of what is indefinitely divisible cannot be safely attributed to Euclid. . . ."[24] Numbers are defined independently of magnitude in the *Elements*. It seems that Euclid wished to keep numbers and magnitude separate.

The fifth Book of Euclid's *Elements*, however, did propose a general theory of ratios

---

[20] Euclid, *Elements*, Vol. II, trans. Sir Thomas Heath (New York: Dover Publication, 1956), p. 277.
[21] *Elements*, Vol. II, p. 277.
[22] *Elements*, Vol. II, p. 278.
[23] For example, VII 14 stated "If there be as many numbers as we please, and others equal to them in multitude, which taken two and two are in the same ratio, they will also be in the same ratio *ex aequali*."
[24] Stillman Drake, "Euclid's Book V from Eudoxus to Dedekind," in *History in Mathematics and Education: Proceedings of a Workshop held at University of Toronto, Canada, July-August 1983* (Paris: Berlin, 1987), p. 53.

and proportionality among magnitudes. This theory is usually credited to Eudoxus of Cnidos, who lived about a half-century before Euclid's *Elements* was written. It is presented in terms of magnitudes only (although magnitudes are never actually defined as they were in Aristotle) and based upon an ingenious concept: 'having a ratio' and 'be[ing] in the same ratio'. Although this theory could have been used for numbers, it was not. In definition III Euclid introduced the idea of **ratio**, "A **ratio** is a sort of relation in respect of size between two magnitudes of the same kind."[25] A ratio is a relation; a relation that is to be considered with respect to quantity, but a ratio is not itself a quantity. Moreover, it is a relation that is only supposed to hold between magnitudes of the same kind. This definition caused many disagreements; Wallis believed that a ratio could be considered a quantity, or even a number, while Barrow insisted that it should actually be considered a relation. Moreover, other early modern commentators such as Ramus believed that the nature of Book V was arithmetic, or algebraic, and they had no doubt that the theory could be applied to numbers.

Definitions IV and V go on to make ratio and proportion theory independent of commensurability, although Euclid never mentions the terms commensurable or incommensurable until Book X. Definition IV stated, "Magnitudes are said to **have a ratio** to one another which are capable, when multiplied, of exceeding one another."[26] In this context, multiplied means multiplied by a number, the definition refers to a "multiple" of the magnitude. The definition does not refer to the product of a magnitude with a magnitude. Multiplication seems to have been the primary focus, because if incommensurability was the most important consideration division would have been necessary,[27] as Euclid's methods for determining incommensurability all required division. This definition also applied to numbers because of the definition of multiply given in VII, Def. 15, and it excludes the concept of an infinitely small quantity that is not zero. It also excludes infinitely large magnitudes because then no finite multiple of the smaller one will exceed the larger. This definition was omitted from the standard Latin versions of the *Elements* until the sixteenth century.[28] Perhaps the first mathematician we will discuss in this study, Robert Recorde, might have studied one of these older versions, but it is difficult to tell as his translation of Euclid did not include Book V.

Definition five provided a method for establishing equality of ratios; it stated

> Magnitudes are said to **be in the same ratio**, the first to the second and the third to the fourth, when if any equimultiples whatever be taken of the first and third, and any equimultiples whatever of the second and fourth, the former equimultiples alike exceed, are alike equal to, or alike fall short of, the later equimultiples respectively taken in corresponding order.

---

[25] *Elements*, Vol. II, p. 114. However I agree with the argument of Stillman Drake, cited above, that it would be more aptly translated as "A *ratio* is a sort of relation in respect of greater or less between two magnitudes of the same kind."

[26] *Elements*, Vol. II., p.114.

[27] Jones, p. 101.

[28] Drake, "Euclid Book V," p. 53.

This definition is pivotal. It says that

$$\frac{a}{b} = \frac{c}{d} \qquad (1)$$

if when we multiply a and c by any positive integer m, say, and b and d by any positive integer n, then for all such choices of m and n

$$ma < nb \text{ implies } mc < nd, \qquad (2)$$
$$ma = nb \text{ implies } mc = nd, \qquad (3)$$

and

$$ma > nb \text{ implies } mc > nd.^{29} \qquad (4)$$

This definition was used in different ways by later mathematicians depending upon whether they believed that these ratios could be treated as quantities, such as numbers, or they believed that the these ratios only described a relation. In general, Book V was considered to be difficult material in the early modern period. Galileo went so far as to dictate an essay on the definitions of Book V in order to make the Euclidean theory easier for students.[30] Book VI went on to apply this theory to plane geometry and to develop a theory of similarity for triangles and parallelograms. It is not until Book X that Euclid first explicitly referred to the concepts of commensurability or rationality. The material in Book X was considered to be even more difficult than the material in Book V in the early modern period. Simon Stevin found that one required a key, in his case a calculus of "surd", or irrational, quantities in order to make the book intelligible.[31] This book began with a series of propositions on ratios of numbers and how they relate to magnitudes. The first definition stated, "Those magnitudes are said to be **commensurable** which are measured by the same measure, and those **incommensurable** which cannot have any common measure."[32] Euclid gave a special status to magnitudes that are commensurable after squaring; he stated in definition two that "Straight lines are **commensurable in square** when the squares on them are measured by the same area, and **incommensurable in square** when the squares on them cannot possibly have any area as a common measure."[33] Numbers, by the above definition coupled with the definition of number are intrinsically commensurable, since they are all collections of units, and they all have a common measure, while commensurability is not inherent in magnitudes. The third and fourth definitions introduce the notions of rational and irrational which makes necessary the introduction

---

[29] Here, *a/b* means the ratio of *a* to *b*, and not *a* divided by *b*.

[30] Stillman Drake, *Galileo at Work: His Scientific Biography* (Chicago: University of Chicago Press, 1978), pp. 422 - 436.

[31] Wilber Knorr, ""La Croix des Mathématiciens": The Euclidean Theory of Irrational Lines," *Bulletin of the American Mathematical Society* 9 (1983): p. 41.

[32] *Elements*, Vol. III., p.10.

[33] *Elements*, Vol. III, p. 10.

of a standard measure. The Euclidean notion of rational, however, differed from our modern usage. In general, his notion of rational seems to have been ignored by early modern authors.

The first proposition of Book X , now known as the "Archimedean property," is the operational definition of continuity, although continuity is not directly mentioned as it is in Aristotle[34]. It stated that

> Two unequal magnitudes being set out, if from the greater there be subtracted a magnitude greater than its half, and from that which is left a magnitude greater than its half, and if this process be repeated continually, there will be left some magnitude which will be less than the lesser magnitude set out.[35]

This proposition utilises Book V definition four, and it therefore links ratios to commensurability. This result also appears in Aristotle's *Physics*. It is used in the next proposition X, 2 which gives the criterion for the incommensurability of two magnitudes, this proposition basically supplies a test. If the procedures from VII 1 and VII 2, the Euclidean algorithm for finding the greatest common measure, are applied to two magnitudes, but fail at each step to measure the magnitude last subtracted, the two magnitudes are incommensurable.

One historian, Wilber Knorr, has labelled Book X "a pedagogical disaster."[36] It was definitely considered to be the most challenging material in the *Elements* by many early modern mathematicians. It did, however, provide definitions and tests for incommensurability that could be utilised by later mathematicians. Indeed, most of the necessary material was covered in the definitions and the first few propositions, and even if the scholars did not follow the entire complex linkage of proofs they could glimpse the necessary procedures.

Overall, although there are some differences between Aristotle and Euclid such as Euclid's seeming to allow proper fractions, they shared many important ideas. It seems more plausible that Euclid presented the material in Book VII, as the Eudoxian proportion theory of Book V could have also served for magnitudes, because the number-magnitude distinction was important to him rather than, as some commentators suggest, because he was merely following a traditional presentation of the material. Charles Jones suggests that Euclid's use of lines to represent numbers might have been the basic source of confusion. Perhaps, since the ratio of positive integers in Book VII are represented as the ratio of lines, it suggested to commentators that numbers in some sense can also be given in terms of lines or magnitudes. Jones states the mistaken idea of some commentators in this way, "That is, since the <u>representation</u> of numbers can be given in terms of magnitudes, then the <u>theory</u> of numbers can be reduced to the theory of magnitudes."[37] But one might also say that as the procedures of Book VII are utilised in Book X, the arithmetic theory of Book VII could have been represented

[34] Jones, "The Concept of One as a Number," p.120
[35] *Elements*, Vol. III, p.14.
[36] Knorr, p. 59.
[37] Jones, p.154.

geometrically because that was the way it would be applied. In general, Euclid preserves the Aristotelian distinction between discrete numbers and continuous magnitudes. Moreover, Euclid's rejection of infinitesimals and his presentation of continuity in X 1 suggest that he also viewed magnitude and continuity in a way similar to Aristotle.

PROCLUS

The first printed Greek text of Proclus' *A Commentary on the First Book of Euclid's Elements* was published in 1533 as an appendix to Simon Grynaeus the elder's edition of Euclid's *Elements*. A Latin translation by Francis Barocius was published at Padua in 1560.[38] Therefore, his book was perhaps available to most of the mathematicians in this study with the exception of Robert Recorde. Isaac Barrow cited Proclus in the context of a discussion of the *Elements*.[39] Moreover, Barrow had one of Proclus' books in his library when he died.[40] John Wallis, in claiming that the ancients had a form of algebra they used as a method of discovery, cited Proclus as an example of a Greek familiar with analysis. Harriot, Napier, and Briggs, however, did not mention Proclus. Nevertheless, Proclus' work was available in this period, however, as we can from Wallis' and Barrow's references to it.

Proclus differed from Aristotle in denying that all mathematical objects are abstracted from sense experience. For Proclus, the objects of mathematical inquiry such as numbers, points, lines, and planes were neither empirical things nor pure forms. He stated, "We must. . . posit the soul as the generatrix of mathematical forms and ideas."[41] Although he followed Aristotle at times by conceiving the mathematical sciences of arithmetic and geometry to be distinct, he sometimes argued that there must be a higher science linking the two areas, a "general mathematics." He stated,

> This science does not consider its province to study the properties that belong intrinsically to numbers, nor to those common to all quantities; rather it contemplates that single form of being or existence that belong to all things, and for this reason it is the most inclusive of all sciences, all of which derive their principles from it.[42]

The idea of a "general mathematics" might have persuaded mathematicians to accept Viète's notion that there could be a higher analysis that dealt with both numbers and magnitudes.

Within the general mathematics, geometry was said to occupy a place second to arithmetic. Proclus agreed with Aristotle that arithmetic is more precise than geometry

---

[38] Proclus, *A Commentary on the First Book of Euclid's Elements*, trans. Glenn Morrow (Princeton: Princeton University Press, 1970), p.xliv.
[39] Isaac Barrow, *The Usefulness of Mathematical Learning Explained and Demonstrated*, trans. John Kirby (London, 1734), p.206
[40] Mordechai Feingold, "Isaac Barrow's Library," *Before Newton: The Life and Times of Isaac Barrow*, ed. Mordechai Feingold (Cambridge: Cambridge University Press, 1990), p.361.
[41] Proclus, p. 11.
[42] Proclus, p.8.

because its principles are simpler, for example, a unit, unlike a point, has no position.[43] For Proclus, the characteristic feature of geometry as distinct from arithmetic was that it deals with incommensurable magnitudes and infinite divisibility. On the other hand, the theory of commensurable magnitudes was developed primarily by arithmetic and then by geometry in imitation of it. He thought that this was why both sciences define commensurable magnitudes as those that have to one another the ratio of a number to a number, and that this implied that commensurability exists primarily in numbers. Proclus stated, "For where there is number there is also commensurability, and where there is commensurability there is also number."[44] This statement would not have encouraged the acceptance of irrational numbers, and it stemmed from Proclus' agreement with Aristotle on the nature of numbers.

The unit was once again claimed to be generative of numbers, or to be the matter underlying numbers, but not to be a number itself. Proclus' position on the possibility of continually creating larger numbers also seems similar to Aristotle's. But instead of using the term "potential" infinite he used the phrase "indefinite increase." Moreover, he pointed out that any number one chose is finite.[45] Much of his discussion of numbers had a Pythagorean outlook; numbers were often depicted as symbols of nonmathematical kinds of being, such as moral virtues or theological and supercosmic realities.[46] In general, when not addressing their mystical qualities, numbers were treated as discrete collections of units. Although in some ways numbers were not considered to be purely abstract, Proclus said "Nor does the student of calculation consider the properties of numbers as such, but of numbers as present in sensible objects; and hence he gives them names from the things being numbered, calling them sheep numbers or cup numbers."[47] Numbers were still, to a certain extent, associated with the objects counted.

Proclus' treatment of magnitude was also similar to Aristotle's. He gave several possible ways to consider lines such as the flowing of a point or magnitude extended in one direction. Yet he was concerned about introducing motion into immovable geometrical objects and how one would move something like a point that has no parts. We are supposed to consider these operations to be taking place in the imagination, "But let us think of this motion not as bodily, but as imaginary, and admit not that things without parts move with bodily motions, but rather that they are subject to the ways of the imagination."[48] Magnitude was still thought to be infinitely divisible, but not into an infinite number of parts.[49] A distinction is made that is comparable to Aristotle's notion of potential and actual infinities, but he did not use this terminology.

While commenting on the problem of bisecting a given straight line, Proclus again

---

[43] Proclus, p.48.
[44] Proclus, p.49.
[45] Proclus, p.5.
[46] Proclus, p. xxvii.
[47] Proclus, p.33.
[48] Proclus, p.146.
[49] Proclus, p.125.

mentions continuity. He clearly states that what is continuous is divisible. Moreover, the continuous consists of parts that are in contact, and these can always be divided. A distinction is made between "is divisible" and "is divisible to infinity." Infinite divisibility must be demonstrated. For example, when geometers demonstrate that there is incommensurability among magnitudes and that not all magnitudes are commensurable with each other, Proclus thought they were then demonstrating that every magnitude is divisible indefinitely and that one can never reach an indivisible part which is the least common measure of a magnitude.[50] These questions led him to consider the problem of the actual infinite.

Proclus believed that Aristotle had already shown that there is no magnitude infinitely extended in any direction. He decided that "the infinite exists only in the imagination." Yet he seemed to view this "imagination" as being somehow inferior to the imagination used to consider a point tracing out a line; he said, "The infinite...is not the object of a knowing imagination, but of an imagination that is uncertain about its object, suspends further thinking, and call infinite all that it abandons, as immeasurable and incomprehensible to all thought. . . ."[51] Although he does not use the same labels, in the end Proclus agreed with Aristotle that the infinite might be in some sense imagined, as when we imagine a larger number or another division of a line, but that the infinite has no actual reality.

Aside from positing a "general mathematics," early modern mathematicians would not have found anything in Proclus on the subject of the nature of numbers and magnitudes that they would not have learned from Euclid and Aristotle. But the idea that arithmetic and geometry had common principles, such as axioms regarding equals and the rules of proportions, might have been seen to be encouraging for a scholar who wanted to develop a theory that would apply to both numbers and magnitudes. Since the soul or imagination was said to be the generator of mathematical forms and ideas, it was not necessary to try to find any sort of physical representation of one's idea. A mathematician need only imagine a one-to-one correspondence between the points on a line and numbers.

---

[50] Proclus, pp.216-217.
[51] Proclus, p. 222.

GREEK ATOMISTS

Modern English algebraists except for Thomas Harriot and a few brief comments made by Barrow were not greatly influenced by atomistic concepts.[52] Additionally, none of the other practitioners discussed in this study other than Harriot and Barrow showed any familiarity with atomistic ideas.[53] However by 1649 Gassendi had published his *Animadversions,* a work that contained a portion of his study of Epicurus. In that same year Walter Charleton published his *Physiologia Epicuro-assendo Charltoniana*, a work widely read in England in the 1650s. Charleton's book permitted atomism to reach a wider audience and it also facilitated the reception of the atomic doctrine into Christian natural philosophy.[54] Of course Aristotle had also provided a summary of early atomist positions. Since it is possible that Aristotle's discussion or one of these latter works influenced some of the mathematicians, a brief summary will be provided of Greek atomistic ideas. Eleatics such as Parmenides and Zeno had argued that the assertion of divisibility of material beings into parts led to unacceptable contradictions and that therefore the real world could contain no plurality or change or motion of any kind. The early atomist responded to this notion by postulating absolutely indivisible and solid pieces of matter that contained no internal differences and were all below the level of visibility. There was also the problem of divisibility at every point; if one believes that a body or line is divisible at every point, and one imagines it so divided: what would be left? Little segments are impossible because then the division would not have been made at every point. But if only sizeless points, or nothing, is left, one is left with the seemingly absurd result that the body or line is made up of points or nothing. It seems that Democritus claimed that atoms were necessary to solve this problem.[55] It must be pointed out, however, that it is doubtful whether the theories of the Greek atomists were mathematical theories at all, and it seems certain that they were not primarily mathematical theories.[56] Moreover, there is no agreement amongst historians as to whether Democritus was a mathematical atomist.

Aristotle in *De caelo* III 4 pointed out that the atomist's denial that division could go beyond a certain point was in conflict with the basic principles of mathematics; the source of the problem is the atomist's commitment to a smallest magnitude.[57] The particular theorem with which the atomist's views differed was "it is possible to take a

---

[52] See chapter four for details of Harriot's atomism and a summary of the hostility towards atomism in England prior to Charleton's work.

[53] Although considering Wallis' involvement with the Royal Society and his knowledge of the science of his times, he might have been familiar with these ideas even though he did not mention them.

[54] Robert Kargon, *DSB*, p.209.

[55] Richard Sorabji, "Atoms and Time Atoms," *Infinity and Continuity in Ancient and Medieval Thought,* ed. by Norman Kretzmann (Ithaca: Cornell University Press, 1982), p.39.

[56] David Furley, *Two Studies in the Greek Atomists* (Princeton: Princeton University Press, 1967), p.5.

[57] Fred D. Miller, "Aristotle against the Atomists," *Infinity and Continuity in Ancient and Medieval Thought*, ed. Norman Kretzmann (Ithaca: Cornell University Press, 1982), p.100.

magnitude smaller than any given magnitude."[58] Therefore, there was at least some discussion of the possible impact of atomism on mathematical practice.

The later atomists had to contend with Aristotle's criticisms of earlier works as well as his theory of continuity. Epicurus (c. 342-270 B.C.) was unable to accept Aristotle's notion of continuity because it involved the idea of potentiality which was in conflict with his fundamental principles. He supposed it was contradictory to say that a finite magnitude is *potentially* divisible to infinity, but to deny that it consisted of an infinite number of parts. For him, to divide was to divide into parts, and a thing consists of those things into which it is divided. Epicurus was "impatient" with the idea of a potentiality that could never be actualised.[59] In order to avoid Aristotle's objections to the theories of the earlier atomists, he proposed the concept of *minima* as indivisibles, not points without magnitude, but units of minimum extension. These *minima* could then be arranged in succession, or side by side, and they would therefore not fall under Aristotle's objection that atoms must touch either whole to whole or part to part. One of the arguments Epicurus gave in support of atomism was that a finite body cannot contain an infinite number of masses because this would make the body infinite in extent. Moreover, he claimed that we must deny that a body can be cut into infinity into ever smaller pieces "in order that we may not make everything weak and be forced in our comprehension of wholes to use up existent things into nonbeing as we shatter them."[60] Epicurus believed that the infinite physical division of a finite body into ever smaller parts would result in the destruction of the body into nonbeing, although he never stated why the infinite division of a finite body would result in nonbeing. He thought that if it is possible for every body to be cut, then it would follow that there would be a time at which all bodies would have been cut, and that this would mean that there would remain no bodies at all. In general, Epicurus' position was similar to the early atomists; there must be ultimate bodies that are indivisible if all bodies are not to change into, or from, nonbeing.

Aristotle had shown that an indivisible body could not move unless time, motion and extension were *all* composed of indivisible units. Epicurus seemed to have accepted that all these things must be composed of indivisibles if any are, but he merely claimed that this was, in fact, the case. He believed in indivisible units of motion and time, such that one unit of motion involves traversing one unit of space in one unit of time. It was never true to say "it is moving," one could only say "it has moved." His indivisibles were ranked in order of the continuum, and their contact was to be considered neither whole with whole nor part with part; they were units of measure that had extension.[61]

Epicurus' and Lucretius' (94 - 55 B.C.) ideas may be quickly summarised as follows. First, they believed that "nothing is generated from the nonexistent. . . ."[62] Moreover, nothing passes away into nothingness. This is basically a notion about the

---

[58] Miller, "Aristotle against the Atomists," p.102.
[59] Furley, *Two Studies*, p.155.
[60] Elizabeth Asmis, *Epicurus' Scientific Method* (Ithaca: Cornell University Press, 1984), p.252.
[61] Furly, *Two Studies*, p. 115.
[62] Epicurus, *Letter to Herodotus*, trans.George Strodach, *The Philosophy of Epicurus* (Northwestern University Press, 1963), p.115.

indestructibility of matter coupled with the idea that the universe is eternal. Second, atoms exist: "The components [of compound bodies] are irreducible and immutable atoms, incapable of decomposition in any manner whatsoever."[63] The totality of "things" was considered to be unlimited, so there were infinitely many atomic bodies and spatial magnitudes as well as infinite space. Lucretius constructed an argument against Aristotle's finite universe. He supposed that a javelin is hurled outward at the edge of an imaginary finite universe: what happens? Either something blocks its flight and prevents it from completing its trajectory or it is borne outward. In the first case, that which blocks it must be something in the space beyond the supposed space, and in the second case, the javelin obviously moves into "outer space". Finally, because of the infinite supply of atoms and space, both Epicurus and Lucretius believed in infinite worlds.

In the chapter discussing early modern English algebra, I will discuss in greater detail the dangers of holding atomic beliefs in the early portion of the seventeenth century. Atomism was associated with Epicureanism and therefore considered to be atheistic (compared with the Christianized Aristotle); it could be used as the grounds for complaint to the secular authorities. In 1591, Robert Parsons denounced Harriot as an Epicurean atheist and conjurer, although it is uncertain whether he was brought to trial at this point.[64] But the stigma of atheism clung to Harriot. The problem was compounded in the period before Charleton's *Physiologia* made atomism more acceptable because Hobbes' notoriety deeply affected the fortunes of atomism in England. He made the paganism of the ancient atomic philosophy a living, burning issue in mid-seventeenth-century England, thus the battle for atomism was begun in the shadow of Hobbes.[65] It is difficult to ascertain exactly how deeply atomistic ideas influenced English mathematicians, since admitting to holding such views would have been risky behaviour in most of the period under discussion.

Although there were several other classical views concerning mathematical continua and the nature of numbers, we have reviewed the concepts actually available to the practitioners in this study. The following sections will discuss the most important of the more modern mathematical developments that influenced the formation of nontraditional views about the relationship between numbers and magnitude and the nature of numbers, and helped to broaden their concept of number.

---

[63] Epicurus, *Letters to Herodotus*, p.116.
[64] Robert Karagon, *Atomism in England from Harriot to Newton* (Oxford: Clarendon Press, 1966), p. 27.
[65] Karagon, p.62.

CHAPTER 3

THE CONTEMPORARY INFLUENCES

This chapter discusses some of the more contemporary developments that might have influenced the progress of early modern ideas about number and magnitude. Since the introduction of mathematical symbolism inspired a broadening of number concepts and the creation of the notion of abstract magnitude, several aspects of the new algebraic methods will be explored. First, we will examine the German cossic notation that was utilised by Robert Recorde, and still included in later algebraic works such as Wallis' *Treatise*. We will also consider the early proto-algebraic techniques of the Italian abacus school writers, since their texts had much in common with Recorde's and they provide us with a demonstration of the changing nature of the problems that practitioners were trying to solve. Second, we shall investigate some of the mathematical contributions of Simon Stevin. His notion of decimal expansion was familiar to Harriot, Napier, Briggs, Barrow, and Wallis and his views about the nature of numbers were diametrically opposed to the Greeks. Third, we will explore the impact of Viète's seminal work that seems to have influenced almost all the practitioners in this study. Descartes' contributions will also be investigated, as Wallis and Barrow closely studied his writings. Although some English mathematicians preferred to attribute Descartes' advances to Harriot, they were nonetheless familiar with his work.

THE ABACUS TRADITION

The abacus school tradition seems to have mainly influenced Robert Recorde. The subsequent algebras of Oughtred and Harriot did not include material of such a practical nature; their works were more theoretically oriented. The abacus school texts, however, do provide us with an idea of the changing nature of the demands made upon mathematics. The source of the abacus school tradition appears to be Arabic mathematics. During the Middle Ages Arabic speaking scholars studied arithmetic and algebra, trigonometry, geometry, and other mathematical and physical sciences as well. Al-Khuwarizmi's algebra, for example, contains a long section on mercantile, geometrical, and recreational problems.[1] It is difficult, however, to trace the exact path of the influence of the Arabic work on the abacus tradition. Clearly, Leonardo of Pisa's *Liber abbaci* of 1202 was the starting point of the European tradition. This work contains almost all the problem types that are

---

[1] Al-Khuwarizmi, *The Algebra of Mohammed Ben Musa,* trans. Frederic Rosen (London: J.L. Cox, 1831)

found in the abacus manuscripts. Pisa quoted Al-Khuwarizmi's work and travelled extensively in Arabic countries. His work was written in Latin and not the vernacular of the abacus manuscripts and a century's time lag intervened between Pisa's work and the first abacus works. Thus, no clear route connects the abacus masters to Arabic mathematicians.[2]

Europe's mercantile development from the thirteenth century onward placed more and more importance on proficiency and understanding of commercial arithmetic. But neither classical mathematics nor the philosophical mathematics taught in universities met the merchants' needs.[3] Classical and university mathematics were out of the merchants' reach partially for linguistic reasons, as many merchants were not equipped to read Greek and Latin. Although Greek mathematical texts were not read in universities, one path of mathematical advancement in the Renaissance was the translation of Greek mathematics texts. The university curriculum in Italy regarding mathematics during the fourteenth and fifteenth centuries was not changed from the medieval curriculum. Boethius' *De arithmetica* was used to teach elementary computation and philosophical number theory. Euclid's *Elements* was used to teach geometry, and John of Sacrobosco's *Algorism* was used to teach arithmetic operations.[4] The approach to arithmetic taken in the universities was focused on a type of "number theory" and the philosophical study of number, unity, equality, ratio and proportion. Things such as perfect numbers fascinated the academics, they looked for order and harmony to apply to metaphysics and theology.[5] The emphasis in the above works was theoretical, but the new economic and social forces were changing the mathematical needs of the merchant community.

The new need for practical mathematics was based upon many things. For example, the reintroduction of a monetary economy and the diversity of monetary systems made advanced mathematics necessary for exchange problems. During the Middle Ages trade had in the main been done through barter. Coinage was not popularly used in the pre-Renaissance period. Thus, no mathematical techniques had been developed to facilitate monetary exchange[6]. A type of chain rule, however, or using the rule of three several times, was developed in the Renaissance in order to facilitate exchange problems. Niccolò Tartaglia (*c.* 1499-1500) felt this aspect of commercial arithmetic to be so important that he devoted many pages of his *General trattato* (pt. II [1556]) to exchange problems. For example, he considered the exchange rates between Venice and fourteen other cities and between Florence and twenty-one cities. Also, partnerships were again coming into use; this made schemes for the correct division of profits a priority. A form of algebra was used to solve these problems. They ranged from the simple division of profit between two partners who both invested capital for the same amount of time to

---

[2] Frank J. Swetz, *Capitalism and Arithmetic* (La Salle: Opencourt, 1987), p.13.

[3] Paul F. Grendler, *Schooling in Renaissance Italy* (Baltimore: John Hopkins University Press, 1989), p.306.

[4] Grendler, p.306.

[5] Grendler, p.306.

[6] Warren van Egmond, "The Commercial Revolution and the Beginnings of Western Mathematics in Renaissance Florence, 1300-1500" diss. Indiana University, 1976, p. 42.

problems with up to five partners in which the partners enter the company at different times. An example of a problem would be:

> There are two partners who make a partnership together and one contributes 300 lire and the other 200 lire and it is to last 3 years and it is agreed between these two that the earnings that they make at the end of three years are to be divided by 1/2 so that (sic) when it comes to the end of 20 months they want to divide and find themselves with earnings of 200 lire. I want to know how much each of them will have?[7]

In order to solve this problem one needs to take into account both the different amount of capital invested by each partner and the length of time of the investment. This requires knowledge of ratios and prorating. A description of the general content of the abacus manuscripts will show that economics and commercial influences led at least indirectly to mathematical advances. Over one hundred Italian folios of handwritten abacus manuscripts have been dated before 1500. They were sewn and bound in leather bindings and produced with care.[8] Great diversity exists between the manuscripts; however, they do share some common elements. All include a collection of problems (which are self-contained units) stating both the problem and the solution. The average number of problems included in the abacus manuscripts was two to three hundred. Apparently, not many problems were duplicated from manuscript to manuscript.[9] Some of the problems are similar in nature to those of the Arabic manuscripts, but differences do exist such as the lack of problems dealing with the correct pricing of slaves. These problems may be classified as belonging to three types: genuine, fictitious and recreational.[10] Recorde's *The Whetstone of Witte* also included these categories. Genuine, or practical, problems are of the type dealing with business or commercial transactions. Fictitious problems are those which are not normally encountered in real life. Recreational problems generally dealt with the properties of numbers themselves. Like Recorde's and Oughtred's works, the abacus texts began with the principles of the Hindu-Arabic number system giving the numerals 0, 1, 2, . . ., 9 and explaining the principle of place value. The only concern for "proof" shown by the abacists was their concern for the accuracy of multiplication and division problems. These "proofs" might more accurately be called checks or confirmations. This concern was perhaps due to the merchants' desire for accurate accounting. Two rules were

---

[7] van Egmond p. 231. The answer is given as: Do it thus: you see first that each of them is to have according to his capital and because one has 2 hundred and the other 3 hundred we add then together, 2 and 3, which makes 5, and 5 will be the divisor. Now we say 2 times 200 makes 400, divided by 5 whence comes 80 and 80 lire will have that which put in 200 and the other should have the remainder of the 200 lire which is 120 lire. Now we say, if they had stayed 3 years and they had these 200 lire of earnings, each would have of it 100 lire. Now you see how much of the 200 lire that one would have above 80 which he is to have by the rule which gives 20 lire because we say in 36 months he would have 20 lire. Now how much would he have had of it in 20? Therefore multiply 20 times 20 which makes 400; divide by 36 whence comes 11 lire 1/9 of a lire, add it to the above 80 and there are 91 lire 1/9 and that man has 91 lire 1/9 of a lire of that 200 lire and the other would have the remainder of the 200 lire, which is 108 lire and 8/9 of one lire. . . .

[8] van Egmond, p. 145.

[9] van Egmond, p.165.

[10] van Egmond, p.165.

used, "proof by seven" and "proof by nine."[11] For example, in "proof by 7", if we say

$$341 * 625 = 213125, \qquad (1)$$

then we divide

$$341 / 7 \qquad (2)$$

with a remainder of 5, then we divide

$$625\ \ 7 \qquad (3)$$

with a remainder of 2, next we multiply the remainders

$$2 * 5 = 10 \text{ and } 10/7 \qquad (4)$$

leaves a remainder of 3, now

$$213125/7 \qquad (5)$$

also leaves a remainder of 3, which helps us to detect incorrect answers. Although this checking procedure has fallen into disuse, the modern standard methods of multiplication and division appear for the first time in abacus books. As with Recorde's work, the genuine problems usually took up the majority of the manuscript.

The business problems covered a wide variety of areas. Determination of the price of a product through use of proportions was typical. Currency exchange problems were popular, in which three values were given and a fourth value was sought. Partnership problems discussed the division of profit according to the amount invested, and variations such as partners entering the agreement at different times. Remnants of the earlier type of economy appeared as barter problems. Usually the concern in these examples was to fix one's profit margins, and thus many of these problems are a mix of cash and barter. The recreational problems, and even some of the business problems that seem to be recreational problems in disguise, exhibit a love of mathematics for its own sake. Many of these problems were of the "find a number such that . . ." type. Also used were "divide a number n into two parts such that the divisors x and $n - x$ were in a specific relationship." For example, "Find me a number such that adding 1/2, 1/4, and 1/5 of it plus three more

---

[11] These rules were based on a principle recognised in the modern theory of arithmetic as the congruence relation of modular arithmetic. For every $a, b \in$ N, $[a] + [b] = [a + b]$ and $[a] * [b] = [a * b]$ where $[a] = a(\text{mod } n)$. This is true because for every $m \in$ N, there exists $n, q, r \in$ N, such that $m = n * q + r$.

makes 25?".[12] In general, some form of "rule" dealt with most problems of all three types.

Sometimes the solutions were worked out purely by logic. No special rules or techniques were applied; they merely reasoned from the given conditions until the answer was found. Many of the interest problems fall under this rubric. The most popular method was the "rule of three" which was the elementary rule of proportionality, if

$$A : B = C : D \text{ then } AD = BC. \tag{6}$$

The pricing and exchange problems used this rule with three knowns and one unknown. The abacus teachers solved many problems this way that we would solve in other ways. Another method was the "rule of simple position" where one made a guess at the solution to the problem and then found the correct answer by means of proportion, using the initial guess, the number that results after its insertion, and the number we want as the three known terms. An example of this type of solution is the problem:

> Find a number of which 5/8 is 29 2/3.
>
> Do it thus: let us propose that this number is 8, and if it were 8, 5/8 of it would be 5, and we have asked that it be 29 2/3. We say this: for the 8 that I proposed I got 5, and I wanted 29 2/3. We must multiply 8 times 29 2/3 giving 237 1/3 and we must divide by 5 giving 47 7/15, and so much is the number.[13]

Our first guess at an answer is 8, which gives us the result 5. These give us the first ratio in the proportion, and the number we want to arrive at, 29 2/3, gives us the third term. Thus,

$$8/5 = x/(\ 29\ 2/3\ ). \tag{7}$$

From Arabic sources the abacists also used the "rule of double false position" when they had two unknowns; again one made guesses at the answers and used the results to find the correct solution. Many examples were given in the manuscripts that illustrated the use of the above methods so that the merchants could gain a feeling for the correct rule to use for the case in question.

Much of the later sixteenth-century algebraic symbolism was abbreviate rather than symbolic of the operations. The term cossic for a writer on algebra derives from the Italian use of the word *cosa*, or thing, to designate the unknown quantity in a problem. Such terms were translations of Arabic algebraical words, for instance, the Arabic word *shai'* is translated as *res* in Latin and *cosa* in Italian. The notation used by Christoff Ruldolff (c. 1525), which was later used by Recorde and included in Wallis' list of algebraic symbols, used the letter *c* with a final loop as a symbol for *cubus*. A sample of the translated terms used are as follows:

---

[12] van Egmond, p. 193.
[13] van Egmond, p. 253.

> *Dragma* or *numerus* is taken here as 1. It is no number, but assigns other numbers of their kind. *Radix* is the side or root of a square. *Zensus*, the third in order, is always a square; it arises from the multiplication of the *radix* into itself.[14]

The *dragma* was, in effect, a symbol that informed one that the number was not attached to an unknown. The *radix* is the symbol for the unknown while the *zensus* is the square of the unknown.

The German, Michael Stifel, in his 1544 *Arithmetica integra* and his other books also used this type of symbolism. Recorde, in his *Whetstone of Witte,* used the symbolism of cossic algebra and format of the abacus texts.[15] Oughtred, on the other hand, used Viète's notation instead of the cossic symbols together with many signs he invented himself, and unlike the abacus texts his work concentrated on problems that did not refer to practical concerns.

Overall, the German and Italian tradition of cossic algebra was similar to the work of the abacus schools in that it operated outside of universities and it concentrated on the solution of practical problems. Because of the nature of its practitioners, much of this type of work displayed no concern with traditional definitions of numbers. It employed an abbreviatory symbolism as we have seen, but it was mainly a large collection of representative problems. These problems were meant to serve as templates for similar problems. Both types of algebra exhibit a problem solving approach to mathematics; the problems they solved were intended to serve the mercantile community, and together with Hindu-Arabic reckoning they provided a means to solve some of the increasingly complex trade and exchange problems.

SIMON STEVIN

Stevin's ideas were familiar to some extent to all the mathematicians in this study except for Robert Recorde. An English translation of Stevin's *De Thiende* by Robert Norton was published in 1608.[16] Napier specifically mentioned Stevin in his *Rabdologia* of 1617, and stated that although he adopted Stevin's principle he would improve Stevin's notation for decimals by introducing the use of a decimal point or comma. Briggs also used Napier's version of Stevin's decimal expansions and used Bombelli's exponential notation in a manner similar to Stevin when he worked with equations. Oughtred does not mention Stevin, or any modern author, by name; he used his own version of decimal notation. The preface of Harriot's *Artis Analyticae Praxis* discusses Stevin's work with algebra as well as his decimal notation. Moreover, Wallis mentioned Stevin's contributions in the introduction of his *Treatise of Algebra*[17] and Barrow directly stated that he agreed with Stevin regarding the acceptability of irrationals and the nature of "Geometrical numbers."[18]

---

[14] Florian Cajori, *A History of Mathematical Notations* (La Salle: Open Court Press, 1974), p. 132.

[15] "Cossike" was the preferred spelling in England in this period.

[16] Simon Stevin, *The Principle Works of Simon Stevin*, Vol. II Mathematics, ed. with commentary by D.J. Struik (Amsterdam: C.V. Swets and Zeitlinger, 1958), p.376.

[17] Wallis in *A Treatise of Algebra,* full citation given below, said "But the first who hath professedly treated of this Subject, and given it the name of *Disme*, or Decimals, (at least the first that I have seen)

There are at least two ways this familiarity with Stevin's work might have influenced early modern ideas about abstract magnitude and the relationship between number and geometrical magnitude. First and most obviously, we must consider the impact of Stevin's decimal notation. Although mathematicians such as Regiomontanus had utilised the notion of decimal fractions before Stevin, they mainly used sexagesimal divisions and they generally only applied this method to trigonometry.[19] Second, Stevin insisted in his 1585 *Arithmétique* that the traditional Greek notion of numbers was wrong: he believed that numbers were continuous rather than discrete.

Stevin's essay on decimal division, titled *De Thiende*, was published in 1585. His work was aimed at the man of practice, thus he wrote in the vernacular and attempted to write as simply and clearly as possible. As previously mentioned, he stated in the preface, To Astronomers, Land-meters, Measurers of Tapestry, Gaugers, Stereometers in general, Money-Masters, and to all Merchants, Simon Stevin wishes health.[20] The notation Stevin developed for expressing decimals was not very successful in that it was quickly superseded. He would place a circle with the number one inscribed inside of it on the right-hand side of the first digit that we would place on the right-hand side of the decimal point, and then next digit in the sequence would have a circle with the number two inscribed inside it and so forth. A circle with a zero inside it denoted an integer. For example,

$$3①7②5③9④ = .3759 = 3759/10000.^{21} \qquad (8)$$

After introducing his notation Stevin explained how to perform simple arithmetic operations and provided practical examples geared towards the professions that he mentioned in his preface. Although *De Thiende* is interesting insofar as it treats numbers as being infinitely expandable Stevin's work on algebra provides more details about his view of numbers.

The *Arithmétique* appeared in 1585, the year his work on decimals was published. It was bound together with another work, the *Pratique d'Arithmétique,* which discussed the commercial applications of the theory in the main sections. Moreover, the book also included Stevin's loose translation of the first four books of Diophantus and a treatise on incommensurable quantities where Stevin attempted to explain Euclid's Book X.

In the section devoted to algebra, Stevin discarded cossic notation and used the exponential notation of Bombelli to label the powers of a given number. For instance,

---

was Simon Stevinus, in a Treatise (which he calls *Disme*) subjoined to his *Arithmetick*, published in *French* . . . ." on page 33.

[18] Isaac Barrow, *The Usefulness of Mathematical Learning Explained and Demonstrated: Being Mathematical Lectures Read in the Publick Schools at the University of Cambridge*, trans. Robert Kirby (London: 1734), p.45.

[19] E.G. Dijksterhuis, *Simon Stevin: Science in the Netherlands around 1600* (Hague: Martinus Nuhoff, 1970), pp. 17-18.

[20] Stevin, *The Principle Works of Simon Stevin*, p.389.

[21] Stevin, *The Principle Works of Simon Stevin*, p.407.

① 2 ② 4 ③ 8 ④ 16 ... (9)

would be the equivalent in our notation of letting $a = 2$, with $a^2 = 4$, $a^3 = 8$, $a^4 = 16$ . . . Interestingly, he allowed the number inside the circle to be fractional.[22] This same circle notation was also used to express "algebraic numbers," or the powers of an indefinite quantity. So 4① + 12 is the equivalent of $4x + 12$. The use of the same notation in three different ways is very awkward and it might have created confusion in the minds of his readers. But despite the problems with Stevin's notation, he made some original claims about the nature of numbers.

Stevin defined number as "that by which the quantity of each thing is explained."[23] Aristotle's dictates were defied in large print; he clearly stated that **one** is a number.[24] Recall that traditionally one had been considered to be not a number at all, but the principle, or beginning of number. Stevin's reasoning was by way of a syllogism: the part of some matter is made of the same matter as the whole, unity is part of a multitude of unities, so unity is of the same matter. Since the matter of a multitude of unities is number, unity is a number. This same reasoning allowed Stevin to accept fractions and irrationals as numbers. Moreover, Stevin insisted that number was not discontinuous quantity.[25] He believed continuous magnitude corresponded to continuous number so the numbers could be considered as a kind of name, or label, for the magnitudes. It appears that the relationship between number and magnitude was supposed to be more complex than the one being a name, or a label, for the other. Stevin stated "To a continuous magnitude corresponds the continuous number to which it is attributed. Number is something **in** magnitude as humidity is in water, it penetrates like this into every part of its magnitude. . . ."[26] Thus number and magnitude were not just the same *kind* of thing, they were, in some sense, *together.*

Euclid's views were also rejected, Stevin stated "An integer is unity or a composite number of unities."[27] Stevin classified numbers as belonging to several different kinds. For example, "arithmetical" numbers were defined as an abstract number "conceived as devoid of magnitude."[28] Geometrical numbers, on the other hand, corresponded to magnitude.[29] This term was used to express his notion that a series of numbers consists of the successive powers and the successive roots of a certain base number. The example we considered was ①2 ②4 ③8 ④16; ① was considered to be a line segment, ② was a square with sides ①, ③ a cube with the same sides, and ④ a parallelepiped.[30] Algebraic numbers were defined in two separate ways. First, an integer algebraic number is a quantity or a composite

[22] Stevin, pp. 460-462.
[23] Stevin, p. 495. "Nombre est cela, par lequel s'explique la quantité de chascune chose."
[24] Stevin, p. 495. "Que l'unite est nombre."
[25] Stevin, p. 501. "QUE NOMBRE N'EST POINCT QUANTITE DISCONTINUE"
[26] Stevin, p. 501. The translation is by Struik, the emphasis is mine.
[27] Stevin, p. 505. "Nombre entier unité, ou composée multitude d'unitez."
[28] Stevin, p. 503.
[29] Stevin, pp. 509-511.
[30] E. J. Dijksterhuis, *The Life and Work of Simon Stevin (1548 - 1620)* (Amsterdam: E.V. Swets and Zeitlinger, 1955), p.23.

multitude of quantities. Second, a fractional algebraic number is a part or consists of parts of integer algebraic numbers. Actually, algebraic numbers may be viewed as multiples of geometric numbers with arbitrary coefficients.[31] Stevin, however, did not believe that any number should be labelled irrational or absurd. He stated, "There are no absurd, irrational, irregular, inexplicable or surd numbers."[32] He believed this to be an error of principle. For instance, $\surd 8$ is not really absurd or irrational, as it can merely be considered a number whose square is 8. Moreover, 8 being incommensurable with its square root did not imply a defect in $\surd 8$. One reason that Stevin might have been so cavalier about breaking with classical traditions about number was his firm belief in a *Wijsentijt*, or age of sages.[33] Although Stevin's belief in a remote past where all things were in perfect order, or a golden age, is usually dismissed by scholars as an odd quirk excusable in a person who contributed so much to science, it might actually have given him the freedom to break with traditional ideas. Even the Greeks, in his view, were lesser men than had existed in the golden age. Hindu-Arabic numerals and algebra were both, for him, remnants of this golden age. Another factor that might have led him to reject the Greek notion of number was his deep commitment to practice. Stevin worked in the fields of mechanics, astronomy, and navigation as well as making contributions to the advance of technology through his work on mills, sluices and locks, military science and architecture. Many of these areas, as Stevin pointed out in his work *La Disme*, benefited from the use of decimal expansions. Whatever Stevin's motives were, his work provided an example of a treatment of numbers as continuous rather than discrete that was available to English mathematicians.

THE TECHNIQUES OF VIÈTE

Viète's work certainly influenced both Oughtred and Harriot. Although Oughtred did not actually mention Viète by name, he adopted Viète's symbolism and took up Viète's idea of a rediscovered analytic art that would open previously closed doors. On the other hand, the introduction of Harriot's *Artis Analyticae* written by Walter Warner, specifically mentions Viète and his influence on Harriot's algebra. Napier's and Briggs' work on the logarithms, however, seems to not have been motivated by the algebraic tradition. Napier's logarithms were developed without the aid of algebraic symbolism of any sort, and Briggs, as has been mentioned earlier, when he used any symbolism at all, preferred the notation developed by Raphael Bombelli that was also used by Stevin. Finally, Wallis and Barrow were both familiar with Viète's analytic art. Wallis was less hostile towards Viète than he was towards Descartes, and he acknowledged him as the creator of the "Specious Arithmetick."[34] But Wallis insisted that Harriot was his primary

---

[31] Stevin, p.519.
[32] Stevin p. 533. "Qu'il y a avcuns nombres absurdes, irrationels, irreguliers, inexplicables, ou sourds."
[33] Dijksterhuis, *The Life and Work*, p.7.
[34] John Wallis, *A Treatise of Algebra, Both Historical and Practical. Shewing, The Original Progress. and Advancement thereof, from time to time; and by what Steps it hath attained to the Height at which it now is* (London: 1685), p.63.

inspiration and while Barrow concentrated on geometry, the symbolism Barrow did use was closer to Harriot's or sometimes Oughtred's.

Viète's methods differed substantially from those of the abacus school or the cossic algebraists. The emphasis was purely theoretical; he did not consider practical problems such as military applications or trade. Indeed, he went as far as giving up the name of algebra altogether. Moreover, in the earlier works symbols only denoted numbers and they played an essentially abbreviatory role. The earlier algebras had been devoted to finding 'recipes' or algorithms that would allow the solution of problems of various sorts, and no general formulas were derived. Viète's famous remark in the introduction to his *Analytic Art* provides us with his opinion of the older techniques, and how he viewed his own contribution:

> Behold, the art which I present is new, but in truth so old, so spoiled and defiled by the barbarians, that I consider it necessary, in order to introduce an entirely new form into it, to think out and publish a new vocabulary....[35]

In this work, Viète established an entirely new form of algebra with new language and methods. Following the work of the abbacus and cossic algebraists, Viète's symbolism arrived after centuries of erratic experimentation with the language of algebra. Even Girolamo Cardano's *Great Art (Ars magna)*, published in 1545, with its famous solution to cubic and quartic equations, was written largely rhetorically (in other words, in prose interspersed with some abbreviations and numerals).[36] Viète boldly moved beyond Cardano in creating symbols that had little or no connection to the entities they represented in that the symbols were arbitrary, chosen without concern for resemblance to the quantities for which they were to stand or words used to express those quantities.

As stated in his preface, in Viète's improved form of symbolism unknowns were distinguished from given magnitudes by being labelled with capital vowels while the given magnitudes were denoted by capital consonants. A calculation carried out using Viète's symbolism might look like this

$$\frac{\text{A cubus - B solido 3}}{\text{C in E quadratum}} \qquad (10)$$

which would be read as "A cubed minus 3 times B solid divided by C times E squared."[37] The equivalent equation in modern notation would be

---

[35] Viète, *Introduction to the Analytic Art*, trans. Reverend J. W. Smith, in J. Klein's *Greek Mathematical Thought and the Origin of Algebra* (New York: Dover Publications, 1992), pp. 318-319.

[36] See Helena M. Pycior's, *Symbols, Impossible Numbers, and Geometric Entanglements: British Algebra though the Commentaries on Newton's Universal Arithmetik*, chapter two "Setting the Scene: The Foundations of Early Modern Algebra" (Cambridge University Press, 1997) for an excellent discussion of both Cardano and Viète.

[37] Jacob Klein, *Greek Mathematical Thought and the Origin of Algebra* (New York: Dover Publications, 1992), p.173.

$$\frac{x^3 - 3b}{cy^2} \tag{11}$$

This can be considered a hybrid of syncopated and symbolic styles in that Viète still used abbreviations to denote powers. Yet this symbolism allowed a more general treatment of problems and it focused attention on the procedures of the solution rather than on the solutions themselves. Michael Mahoney has claimed that the introduction of wholly symbolic solutions to general equations suggested that the solutions could be applied to other than numerical problems such as geometrical line segments.[38] Indeed, the algebra of Viète was designed to be as much geometric as arithmetical, and he intended his procedures to be entirely general.

Analysis, for Viète, was "taking the thing sought as granted and proceeding by means of what follows to a truth that is uncontested."[39] He separated analysis into three different kinds: zetetic, poristic and exegetic. The first two types of analysis he believed to have been discovered by the ancients and he only claimed to be the originator of the third kind. Algebra, or, for Viète, the "threefold analytic art" or "the science of right finding in mathematics," was composed of these three forms. For Viète, the zetetic art was the method "by which is found the equation or proportion between the magnitude being sought and those that are given," the poristic art was the method "by which from the equation or proportion the truth of the theorem set up is investigated," and the exegetic art was the method "by which from the equation set up or the proportion there is produced the magnitude itself which is being sought."[40]

Viète was influenced by the classical beliefs that have been previously discussed into holding that only magnitudes of "like genus" could be compared. He called his law of homogeneity "the supreme and everlasting law of equations or proportions" and it stated that "only homogeneous magnitudes are to be compared with one another."[41] The Greeks, he noted, had said that it is impossible to know how heterogeneous magnitudes could be "cojoined." Numbers, of course, have no dimension. Thus the product of any combinatory operation on a number is a number, so numbers presented no difficulties in this regard. Viète, however, accepted the Greek notion that the product, for example, of two line segments would be the rectangle constructed of lines as adjacent sides. Thus, every equation in Viète's "analytic art" had a dimension, and the dimension was directly connected to the degree of the equation, but he did not ignore equations of the fourth degree or higher as having no meaning.

There are several ways, as we have seen, that Viète's "analytic art" might have influenced the development of early modern English mathematics. Viète avoided the issues of the relation between geometric magnitude and numbers by defining the

[38] Michael Mahoney, *The Mathematical Career of Pierre De Fermat (1601 - 1665)* (Princeton University Press, 1973), p.35.
[39] Viète, p.320.
[40] Viète, p.321.
[41] Viète, p. 324

algebraic operations independently of whether they applied to number or non-numerical quantities (such as line segments or other geometrical magnitudes). As Bos puts it, Viète "legitimized the use of algebra in geometry while retaining the conviction that geometrical magnitude were essentially different.[42] However, as Klein suggests, his methods could be viewed as being designed to apply equally to numbers and geometrical magnitudes. Thus Viète's work was a universal method for solving problems. The abacus and cossic texts had dealt entirely with numbers, which is perhaps not remarkable given the nature of the problems they were solving. But Viète's techniques, applicable to both numbers and geometric magnitudes, helped to create the notion of abstract magnitude which would have been less obvious using the older methods. Second, Viète's created symbolic techniques of calculation that could be applied to magnitudes of different natures be they number, geometrical magnitude or otherwise. These mathematical entities were to be considered only in their aspect of being a magnitude abstracted from their actual nature. Viète's algebra dealt with abstract magnitudes symbolically represented by letters.

Third, the new methods led to a study of relations that pointed the way to being able to solve problems in a more general way than ever before. The older abacus and cossic techniques were closer to recipes or trial and error methods. The underlying similarities between the problems were hidden from view. Yet Viète remained somewhat attached to the traditional Greek distinctions and a geometrical way of thinking as can be seen in his insistence on homogeneity. It was not until Descartes that this constraint was thrown off and a new relationship was invented for magnitudes of different degrees as well as number

DESCARTES

None of the mathematicians in this study except for Wallis and Barrow were familiar with Descartes' work. Wallis' attitude towards Descartes was quite hostile. He stated, "In sum, He [Harriot] hath taught (in a manner) all that which hath since passed for Cartesian method of *Algebra*, there being scarce any thing of (pure) Algebra in DesCartes, which was not before in Harriot; from whom Descartes seems to have taken what he hath. . .but without naming him."[43] Oughtred was credited along with Harriot as being Wallis' primary influences, and Descartes was given scant praise.[44] Wallis went as far as saying that Harriot had made "very many advantagious improvements in this Art; and hath laid the foundations on which DesCartes ...hath built the greatest part (if not the whole) of his *Algebra* or Geometry. Without which, that whole Superstructure of DesCartes (I doubt) had never been."[45] Wallis' hostility towards Descartes seems to have developed later in his life since in the *Arithmetica Infinitorum* (1656) he ranked him with the greatest

[42] Bos, *Redefining Numerical Exactness*, p. 146.
[43] Wallis, *Treatise*, introduction.
[44] Wallis said that he read Oughtred's *Clavis* in 1649, and that this reading motivated him to attempt to find original mathematical methods. *Treatise*, p.121.
[45] Wallis, *Treatise*, p.126.

mathematicians of the era.[46] But whatever the source of Wallis' animosity, he was certainly familiar with Descartes' work, even if he would not admit to being influenced by it.

Barrow, on the other hand, admired Descartes' mathematical achievements. In his *Mathematical Lectures* he called Descartes' methods "most elegant".[47] As early as 1648, John Smith lectured on Descartes' *Geometry,* and by the 1660s Descartes' ideas were being widely discussed at Cambridge.[48] Barrow is said to have studied Descartes' works while still an undergraduate.[49] In 1652, in an oration exercise towards his M.A. degree, Barrow said of Descartes that he was "extraordinary skill[ed] as a mathematician."[50] Yet Barrow was concerned about Descartes' neglect of an experimental study of nature, in contrast with Bacon and Gilbert, and "the grave religious implications of Cartesianism rooted in the banishment of the spiritual and immaterial."[51] Since both Wallis and Barrow were acquainted with Descartes' work, it will be necessary to point out how it differs from Viète's, although it would take us too far from our topic to explore it in the detail it deserves.

In the *Regulae*, rule XIV tells us that continuous undivided magnitudes, such as lines and planes, permit us to formulate their proportions (and therefore their equations) only when their common measure, or unit, is known. Once, however, this condition is met, these continuous magnitudes can be understood as numbers. Descartes said, "We must also understand that continuous magnitudes can, by the help of the unit which has been assumed, sometimes be reduced entirely to multitude."[52] Klein believes that for Descartes figures represented not a determinate number of units of measurement but the "symbol of an indeterminate multitude obtained by symbol-generating abstraction."[53] In other words, they represented abstract magnitude. Moreover, Klein claims that it is this "symbolic" character of the figures that makes possible the what he believed to be the mutual correspondence of lines and numbers in Cartesian mathematics. In some ways Descartes' emphasis on symbolism was not unlike Viète's, but unlike Viète, Descartes eventually discarded the "Law of Homogeneity" that made operations on numbers different in some ways from operations on magnitudes.

There were two parts to Descartes' method in his *Géométrie*. First, there was an analytic part that used algebra to reduce any problem to an appropriate equation. Second, there was a synthetic part that found the appropriate construction of the problem on the basis of the equation. In order to translate problems into algebraic equations one must interpret operations such as addition and multiplication in such

---

[46] J. F. Scott, "John Wallis as a Historian of Mathematics," *Ann. of Sci.* Vol. 1, no. 3., p. 337.
[47] Barrow, *The Usefulness of mathematical Learning*, p. 183.
[48] Feingold, "Isaac Barrow," p. 25.
[49] Abraham Hill, "Some Account of the Life of Dr. Isaac Barrow," in *Theological Works of Isaac Barrow*, ed. Alexander Napier, 1: xxxvii-liv, pp. xl-xli.
[50] Feingold, "Isaac Barrow," p. 28.
[51] Feingold, "Isaac Barrow," p.29.
[52] Jacob Klein, *Greek Mathematical Thought and the Origin of Algebra*, p. 204.
[53] Klein, p. 205.

as way as to make them applicable also to geometrical objects such as line segments.[54]

Descartes rejected the notion of homogeneity in the first book of his *Géométrie*, which opens with the claim:

> Any problem in geometry can easily be reduced to such terms that a knowledge of the lengths of certain straight lines is sufficient for its construction.[55]

Viète and Descartes had both noticed that the successive powers of the unknowns form a continued proportion

$$x : x^2 = x^2 : x^3 = x^3 \ldots \tag{12}$$

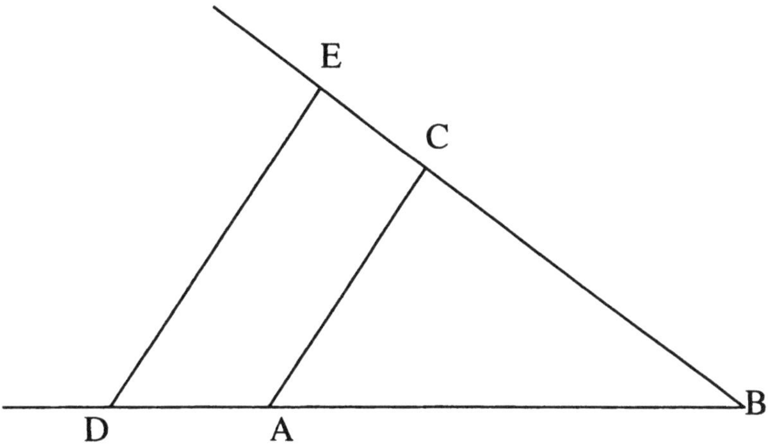

Figure 1.

Descartes, however, surpassed Viète by observing that the proportion began with a unit magnitude. He used this notion to show "how the arithmetical calculus is related to the operations of geometry." If we let *a* and *b* be line segments, then their addition and subtraction are unproblematic even with homogeneity constraints. In order to preserve homogeneity in multiplication, he designated a unit magnitude to which all other unit magnitudes could refer. So *ab* can be constructed by letting AB = 1 be the unit segment, and assuming that we want to multiply the segment BD (or

[54] Bos, *Redefining Geometrical Exactness* includes an extended treatment of elements of Descartes' mathematical practice.

[55] Rene Descartes, *Discourse on Method, Optics, Geometry, and Meteorology*, trans. Paul Olscamp (Indianapolis: Bobbs-Merrill Co., 1965), p.177.

a) by BC (or *b*). We can do this by joining A and C, and then extending the line BC indefinitely, and drawing a line parallel to AC, with one endpoint being D, and the other endpoint E being where the line being drawn intersects the extended BC. Then BE = BD • BC = *ab*.[56] It is simple to check that this is correct by using the proportionality between triangles ABC and DBE. Descartes developed similar constructions for a/b and $\sqrt{a}$, and he also introduced the notation a^2, a^3 and so on for powers of a.[57]

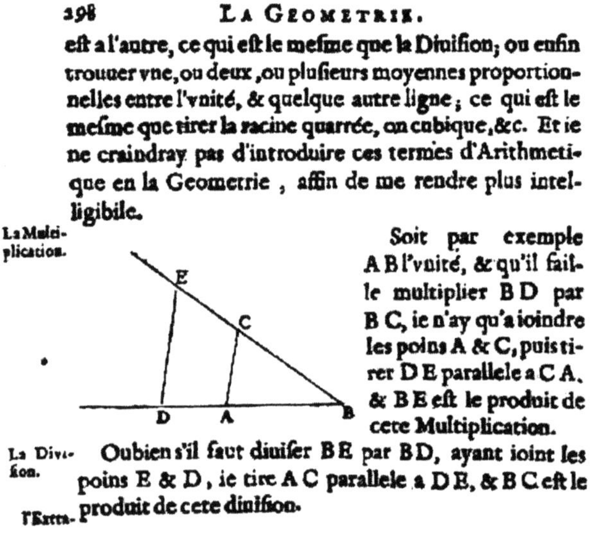
298 LA GEOMETRIE.

est a l'autre, ce qui est le mesme que la Diuision; ou enfin trouuer vne, ou deux, ou plusieurs moyennes proportionnelles entre l'vnité, & quelque autre ligne; ce qui est le mesme que tirer la racine quarrée, ou cubique, &c. Et ie ne craindray pas d'introduire ces termes d'Arithmetique en la Geometrie, affin de me rendre plus intelligibile.

La Multiplication.

Soit par exemple A B l'vnité, & qu'il faille multiplier B D par B C, ie n'ay qu'a ioindre les poins A & C, puis tirer D E parallele a C A, & B E est le produit de cete Multiplication.

La Diuision.

Oubien s'il faut diuiser B E par B D, ayant ioint les poins E & D, ie tire A C parallele a D E, & B C est le produit de cete diuision.

l'Extra-

Figure 2. Descartes' Diagram

Descartes' new interpretation allowed him to adopt a general strategy for solving all geometrical problems. He proposed that when one wanted to solve some problem, that one should consider it solved, and give names to all the lines, unknown as well as known, that might be needed to solve the problem. Next, he thought one should ignore the difference between the known and unknown lines, and find the most natural relation that holds between the lines. In this way, one would arrive at one (or several) equations.[58] Finally, the equation must then be constructed, or in other words, its roots must be found geometrically. Considering only those problems which can be constructed by a ruler and compass, then one will be led to a second-degree equation and so one must construct the root of such an equation.[59]

Let us consider, for example, the construction of the positive root of the equation

---

[56] Descartes, pp. 177-178.

[57] Descartes, p. 180.

[58] Descartes said " And we must find as many such equations as we assume there to be unknown lines." p. 179.

[59] Descartes, pp. 179-180.

$$z^2 = az + b^2 \tag{13}$$

with a and b positive. To construct z, he created the right triangle NLM, whose side LM is equal to b while the other side LN is equal to $1/2\ a$. Then he extended MN, the base of the triangle, to O, such that NO is equal to NL. Then OM is z, the required line.[60]

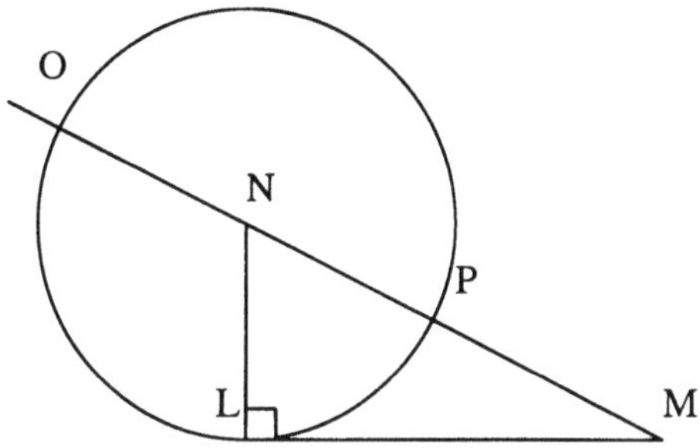

Figure 3.

We know from the Pythagorean theorem that

$$MN^2 - NL^2 = LM^2, \tag{14}$$

and because

$$MN = OM - NL, \tag{15}$$

and by substitution

$$(OM - NL)^2 - NL^2 = LM^2 \text{ or } OM(OM - 2NL) = LM^2. \tag{16}$$

But we had $2NL = a$ and $LM^2 = b^2$., as $z = OM$, we get $z(z - a) = b^2$ or

$$z^2 = az + b^2.\text{[61]} \tag{17}$$

[60] Descartes, p.181.

[61] Paolo Mancosu, "Descartes's *Géométrie* and Revolutions in Mathematics," in *Revolutions in Mathematics*, ed. Donald Gilles (Oxford: Clarendon Press, 1992) contains an extensive summary of the highlights of this text. See also, Paolo Mancosu's *Philosophy of Mathematics and Mathematical Practice in the Seventeenth Century* (Oxford: Oxford University Press, 1996), pp. 65-91.

Comment ils ſe reſoluent.

Et lors cete racine, ou ligne inconnue ſe trouue ayſement. Car ſi i'ay par exemple

$z^2 \propto a z + bb$

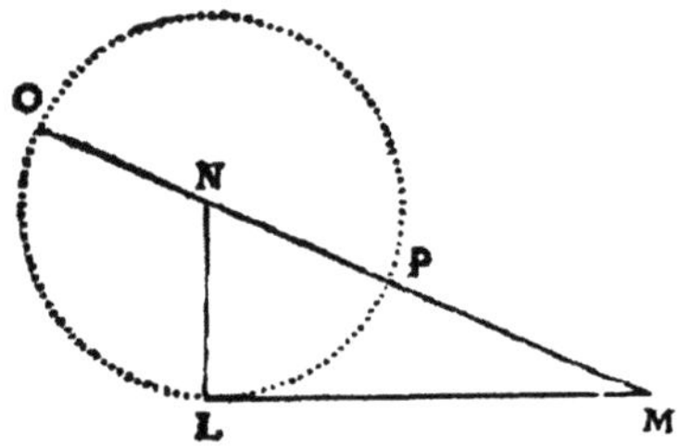

ie fais le triangle rectangle N L M, dont le coſté L M eſt eſgal à *b* racine quarrée de la quantité connue *bb*, & l'autre L N eſt $\frac{1}{2} a$, la moitié de l'autre quantité connue, qui eſtoit multipliée par *z* que ie ſuppoſe eſtre la ligne inconnue. puis prolongeant M N la baze de ce triangle,

Figure 4.

In this way Descartes was able to apply algebraic techniques to the solution of geometric problems without concerning himself with homogeneity constraints. However, H. J. M. Bos has pointed out in his article, "On the Representation of Curves in Descartes' *Géométrie,*" that there is a tension "between geometrical and algebraic methods of definition and criteria of acceptability" in this work.[62] Although algebra became Descartes' primary tool, both for the solution of problems and the classification of curves, he still believed in the principle of geometrical construction by means of curves traceable by instruments. In spite of this conflict between methods, Descartes' work strengthens the association between unknowns and line segments. Descartes' interpretation of algebraic operation as applied in geometry involved geometrical magnitudes, line segments, not numbers. Descartes did not identify line segments with their numerically expressed lengths. Rational and irrational number used as factors, terms, or solutions to equations in the *Géométrie*. As Bos pointes out, "most of these numbers were rational and stood for rational scaler factors or for line segments defined as rational multiples of the unit."[63]

A complete correspondence between algebra and geometry was not achieved at the beginning of the text. Descartes stated, "I say nothing about the cube root, nor about the others, because it will be more connivent to deal with them later." Higher-order roots could not in general be constructed with straight lines and circles, and Descartes delayed this presentation until the final chapter of his book.

[62] H. J. M. Bos, "On the representation of Curves in Descartes' *Géométrie,*" *AHES* 24 (1981) p.278.
[63] Bos, *Redefining Geometrical Exactness*, p. 296. I am following Bos' interpretation of Descatres' text here.

CONCLUSION

Early modern English and Scottish mathematicians had a variety of sources of inspiration for their ideas of numbers, geometrical magnitude, and abstract magnitude. The practitioners in this study were familiar with traditional Greek notions of number and geometric magnitude.[64] Moreover, they also had access to certain modern developments that might have encouraged them to partially reject the classical distinction between continuous magnitude and discrete number. New types of practical problems whose answers could not be expressed in terms of positive integers were one factor that led to the changing views on the relationship between numbers and magnitude. Another consideration was the evolution of a symbolism for unknowns that eventually were used interchangeably on both numbers and magnitudes. Both Viète's, and to an even greater extent Descartes', methods could seen as introducing the notion of abstract magnitude. Stevin made available a technique to express irrational numbers and fractions to the desired degree of accuracy, and improved the ease of numerical operations with such numbers. Almost every mathematician in this study utilised decimal expansions no matter what their views were on the nature of numbers. Stevin also provided an example by his vehement rejection of the Greek separation of numbers and magnitude. Now that we have summarised the necessary background material, we can turn to the practice of the early modern algebraists to explore what the English practitioners actually did with these notions. We will discover how the changing nature of the problems they were solving coupled with the evolution of mathematical symbolism influenced their ideas about the relationship between number and magnitude

[64] The exception might be Napier; it is difficult to tell if he was acquainted with the Greek ideas but he was certainly influenced by certain contemporary ideas, such as Stevin's.

CHAPTER 4

EARLY MODERN ENGLISH ALGEBRA

A review of several early modern English algebras will provide a glimpse of the beginnings of numbers being treated as continuous in the Aristotelian sense rather than discrete. Some of these texts claimed to hold to the classical notion of numbers, but they also operated on fractions and irrationals without committing themselves as to their actual status as numbers. These algebras also show a progression from abacus style, or problem oriented texts, to a purely symbolic, abstract style. Although this chapter traces this development of English algebra, it is not our primary focus. Instead, we will investigate whether the shift from a practical to a theoretical orientation influenced the treatment of numbers in algebraic texts.

The relationship between theoretical and practical mathematics was already quite close in England in the late 1500s due to a combination of government interest in practical mathematical applications (ballistics, navigation, and surveying) and increased mercantile activity. A host of new problems with mathematical solutions were introduced into England from the late sixteenth century: naval warfare, expanding trade, more intricate and sophisticated merchants' accounts, the needs of customs officials, excise men, gaugers, and assayers all demanded more complicated numerical solutions.

Another factor that caused numbers to be treated as though they were continuous, in the sense that fractions and irrationals were accepted as solutions to equations, was the maturation of the algebraic methods themselves. The texts we study in this chapter document the evolution from an abbreviatory type of symbolism in which each power of an unknown required a separate symbol to a totally general symbolism in the style of Viète. This improved notation encouraged the notion of abstract magnitude. An unknown could act as a symbol standing for either a number or a magnitude, and this fact weakened the distinction made by the Greeks between discrete number and continuous magnitude.

Several early modern algebraic texts will be investigated in order to discover how algebraic methods and choice of problems influenced numeric representations. The first text, Robert Recorde's (1510-58) *The Whetstone of Witte*, published in 1557, includes practical problems whose answers could not be expressed in terms of positive integers. This work was the first English algebra, and it will provide a framework for our discussion of the later texts in that it shows the level of algebraic knowledge available when these techniques were first introduced into England. For the sake of comparison, Humfrey Baker's (fl. 1557-87) *The Well Springe of Sciences* (1562), and Thomas Blundeville's (fl. 1560-1602) *Exercises Containing Sixe Treatises* (1594), texts that include arithmetic, but no algebra will also be investigated. One of the few English texts written between 1557 and 1631 that include any algebraic material, John Tapp's (fl. 1596-1631) *The Path-Way to Knowledge* (1613), will also be briefly examined. Finally, William Oughtred's (1575-1660) *Clavis Mathematicae* (1631) and Thomas Harriot's (1560-1621) *Artis Analyticae Praxis* (1631), books typical of English algebra in a more mature phase, will also be explored.

THE NATURE OF THE TEXTS

Although both Recorde's and Tapp's texts employed symbols, they were still mainly rhetorical and informally presented in the form of a dialogue. Baker and Blundeville's style was a mixture of definitions with questions and answers. Oughtred and Harriot's works, on the other hand, were formal mathematics texts and the problems were expressed mainly by means of symbols and equations. These texts taken together exemplify both factors we wish to investigate: the impact of the changing nature of the problems and the development of more sophisticated algebraic techniques.

The reason for investigating texts so widely separated in time as Recorde's, Tapp's, Baker's, Blundeville's, Oughtred's and Harriot's is that there was a dearth of algebraic material published in England between 1557 and 1631. Even in Tapp's work, the algebraic section took up less than one fourth of the text. Books dealing mainly with arithmetic, such as Humfrey Baker's *The Well Springe of Sciences* or Thomas Masterson's *The First Book of Arithmetic* met the needs of the market for mathematical textbooks without introducing algebraic methods. Thomas Blundeville, in his 1594 *Exercises, Containing Sixe Treatises*, mentions that he is not including any algebra because he hopes an unnamed friend would soon publish a book on this topic.[1] Although Leonard Digges (1510-58) wrote a book, published by his son Thomas (circ. 1543-95) in 1572, titled *Stratioticos,* that included some algebra, it was geared towards military uses. The authors merely included "so much of the rules and equations algebraical and art of numbers cossical, as are required for the profession of a soldier"; the algebraic section was a very small part of their text.

There are, however, difficulties in placing Recorde, Tapp, Oughtred, and Harriot together as algebraists. Although they pursued related mathematical methods, they came from quite different traditions, and had different motivations. Recorde dedicated the first English algebra, *The Whetstone of Witte,* to the Muscovy Company in November 1557, and his earlier *Castle of Knowledge* (1556), a treatise on the sphere, was written expressly for the Muscovy Company's navigators. The Muscovy Company's goal was to find an alternative route to the East that would avoid conflict with the Iberian powers.[2] Increasing trade created new possibilities for patronage; the tonnage of shipping doubled between 1570 and 1630, and by the end of this period the English merchants and ships handled a complex network of

---

[1]Thomas Blundeville, *His Exercises, Containing Sixe Treatises*... (London, 1594) Sig. A4.

[2] Kenneth R. Andrews, *Trade, Plunder and Settlemen* (Cambridge, Cambridge University Press,1984), p. 10.

trade.[3] John Tapp tried to appeal to a similar audience. Transforming himself from a member of the Draper's company into a teacher of navigation, he joined the Stationer's Company and began to publish nautical books and mathematical works that would appeal to mariners and other practical men.[4] Tapp began by dedicating his *Path-Way to Knovvledge* (1613) to "Sir Thomas Smith, Knight, Gouernor of the Company of Marchants of London Trading the East-Indies and the Moscouie Company, as also the Company of Discouerers for the North-West passage, and treasure for the plantation in Virginia," making it clear that he was trying to attract an audience for his mathematics whose primary interest was navigation. Smith was one of the greatest promoters of overseas enterprises, and one of the prime undertakers for the discovery of the North-West passage. [5] Tapp appealed to this audience to learn mathematics because it "is the chief and most effectuall branch" for the "nourishment" of navigation. He praised his patron for providing ships, mariners, settled trade and new discoveries " . . . to the great benefit of many thousands imployed therein, a continvall profitte to the common-wealth, and a lasting glory to our Nation."[6]

Oughtred's and Harriot's works were directed towards a different type of audience. Oughtred did not depend upon income from either his students, texts, or instruments. He was ordained in 1603, had the living of Shalford, Surrey, and later was the rector of Albury until his death.[7] At most, his students supplied him with an admiring audience and occasional presents. As Oughtred did not have to attract paying students he could afford to concentrate upon his own vision of mathematics, as can be seen in the text professedly created for William Howard's use, which appeared in 1631 with the title *Arithmeticae in Numeris et Speciebus Institutio: Quae tum Logisticae, tum Analyticae, atque adeo Totius Mathematicae quasi Clavis Est*. Several other editions in Latin, titled *Clavis Mathematica*, and in English, *The Key of the Mathematics New Forged and Filed*, were published throughout the century.[8] The *Clavis* contained, in a condensed and intensively symbolic form, the essentials of arithmetic and algebra known at that time, and neither Oughtred's language nor his choice of material was aimed at the general public. And, as we will see, Harriot's work, compiled out of his mathematical papers after his death and published in Latin, was intended more as a tribute to his mathematical ability than as a textbook for aiding the general public to master algebraic techniques.

---

[3] Kenneth Andrews, *Trade, plunder and settlement: Maritime enterprise and the genesis of the British Empire 1480-1630*, (Cambridge, 1984), p. 8.
[4] E. G. R. Taylor, *The Mathematical Practitioners of Tudor & Stuart England*, (Cambridge, 1954), p. 193.
[5] Kenneth Andrews, *Trade, plunder and settlement: Maritime enterprise and the genesis of the British Empire 1480-1630*, p.20.
[6] John Tapp, *The Path-Way to Knovvledge* . . ., (London, 1613), Sig. A2[r].
[7] See Frances Willmoth's excellent *Sir Jonas Moore: Practical Mathematics and Restoration Science* (Woodbridge: Boydell Press, 1993), p. 45, for Oughtred's early patrons among the local Surrey gentry.
[8] The *Clavis* appeared in five Latin editions, the first in 1631, and the last in 1698. There were also two independent English editions. The first English edition, London, 1647, was translated by Robert Wood.

ROBERT RECORDE

Recorde's text also differs from the later work of Oughtred and Harriot in that it uses German cossic notation, and contains detailed sections of practical examples. His methods were partially derived from such German sources as Johann Scheubel and Michael Stifel.[9] Recorde's content is similar in some respects to the previously discussed material taught in the Italian abacus schools. Some of these books used equations and unknowns although the problems were mainly considered rhetorically. Recorde's work shared the personal, conversational style of the abacus texts. The similarities between Recorde's work and the earlier work of the Italian abacus masters might in part be explained by the backwardness of England during this period. England lagged behind the Continent not only in mathematical theory, but also in areas that might have supported mathematical practice, such as instrument making.[10]

Recorde differs from the abacus teachers insofar as he received a B.A. at Oxford (1531) and a M.D. at Cambridge (1545), while abacists did not usually obtain a university education.[11] There is little evidence of a link between abacists and humanists, and Recorde is known to have had humanist sympathies. Recorde, who for a time practiced as a mathematics teacher and a physician, was the first comptroller of the Bristol Mint (1549), and later General Surveyor of the Mines and Monies in Ireland (1551). He wrote mathematics texts that were blends of theoretical and practical topics laced with instrumental operations.[12] In general, his status was higher than the typical abacus style teacher.

Also, Recorde's invention of the "=" sign made his algebra totally symbolic, unlike earlier abacus works. The *Whetstone*, however, seems to mainly follow the practical tradition of the cossic algebraists and the abacus schools. Perhaps these differences are partially due to the lateness of the introduction of the algebraic and the new mercantile techniques into England. It was not until 1553 that the Muscovy Company was founded, with the Merchant Adventurers following in 1564.[13] We must keep in mind that Tudor England remained technically backward in economic matters compared with the best practices of continental Europe. It was not until the middle of the sixteenth century that England started to catch up.[14]

---

[9] Joy Easton, *Dictionary of Scientific Biograhpy*, p. 339.

[10] A.J. Turner, "William Oughtred, Richard Delamain and the Horizontal Instrument in Seventeenth Century England," in *Annali Dell'Istituto e Museo di Storia della Scienza di Firenze* 6 (1981), p.101.

[11] Warren Van Egmond, "*The Commercial Revolution and the Beginnings of Western Mathematics in Renaissance Florence, 1300-1500*", diss. Indiana University, 1976, p.121.

[12] *The Grounde of Artes* first appeared in 1543, and forty-five editions of this work were printed under various editors; *The Pathewway to Knowledge*, London, 1551 (also in 1574 and 1620); *The Castle of Knowledge*, London, 1556 (reprinted 1596); *The Wherstone of Witte*, London, 1557. For fuller details of Recorde's life, see: F.M. Clark, "New Light on Robert Recorde", *Isis* (1926), **8**, 50-70; E. Kaplan, "Robert Recorde: Studies in the Life of a Tudor Scientist", PhD dissertation, New York University, 1960.

[13] D.C. Coleman, *The Economy of England 1450-1750* (London: Oxford University Press, 1977), p.59.

[14] Coleman, *The Economy of England 1450-1750*, p.69.

Recorde's chief aim was to kindle interest in the many useful applications of mathematical knowledge; he wanted his readers to be able to employ their new learning to solve specific problems. He was writing for the beginner and not for the learned. He first gave a clear exposition of the fundamental facts, and the proof or demonstration was left until the student had mastered the basic concept

The Whetstone of Witte was written in the form of a dialogue between a master and his scholar, or student. Early in this work Recorde shows his adherence to the traditional Aristotelian conception of number. The Master begins by stating "Unitie is of it self indivisible," and that unity is the origin of number.[15] Recorde echoed Aristotle by having the printers set a note in the margin of the text that "One is no nomber."[16] It thus appears that Recorde accepted the classical Aristotelian-Euclidean notion of number being collections of discrete units. On the other hand, the necessities of practice seem to have created a certain amount of tension concerning number concepts in Recorde's work. He acknowledged that earlier writers such as Euclid and Boethius only recognised "whole numbers," (positive integers) as numbers. He was, however, willing to work with the positive rationals, which he called "broken" numbers, although he was uncertain about their status as numbers. The Master claimed that all numbers consisted of a multiple of units, and since fractions were less than units, they could not "exactly be called nombers"; yet he still showed how one could perform arithmetic operations with "broken nombers".[17] The solutions to many of his examples required, at the very least, the positive rationals; Recorde solved this difficulty by broadening his number concept while simultaneously using labels that signalled his awareness of the unacceptability, by traditional standards, of the new numbers.

Numbers also were classified by Recorde as being "figuralle," or as representing some figure. He carefully gave diagrams and examples of squared and cubed numbers, then used this idea as an aid to finding the roots of numbers. For example, he states that "2 tymes 2 doeth make 4," which is a square number, and an example of a cubed number is "2 tymes 2 twise, maketh 8." In other words, in modern terms, for any integer n, n^2 would be a "square" number and n^3 would be a "cubed" number.

---

[15]Robert Recorde, *The Whetstone of Witte,* (London, 1557), sig. Ai^{v}.
[16]Recorde, *Whetstone of Witte*, sig. Aiii^{r}.
[17]Recorde, *Whetstone of Witte*, sig. Aii^{v}.

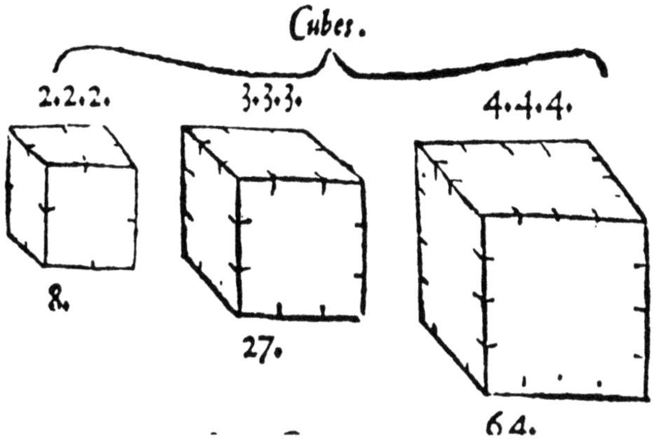

Figure 1[18]

Like Aristotle, he viewed this multiplication process as potentially endless, although one could not, for example, draw a figure for n^4. He provided a short list of names of the potential multiplication of numbers that included geometric figures as in figure 1 as an aid to understanding.

Recorde also had an inkling of the possibility that some numbers could be represented to any desired degree of accuracy. He believed that surd numbers[19], for example, could be approximated with fractions as closely as one desired, and he seemed to appreciate that this process could go on indefinitely. The labelling, however, of the broadened number concept still created tensions. In his section dealing with the extraction of roots, he noted that the root will not always be a number in his usual restrictive sense. Recorde stated that his usual method of extracting roots only served for numbers that "bee fully square", recognising that "Other nombers there bee infinite, whiche be not square, and therefore have no square rootes. Yet often tymes it happeneth, that we shall bee occasioned to search for the nigheste nomber, that maie resemble their rootes."[20] He showed how one can compute roots of such numbers to the desired degree of exactness by representing them partially with "broken nombers". He gave as an example the taking the root of 296,882. Using his normal procedure, he calculated an integer close to the "true" root, obtaining 544. At this point he noted that if one squares 544, the result is 295,936, which misses the target by 946. Recorde stated that "some men" tried to achieve a closer fit at this stage by adding one to the integer approximation, noting that this attempt at correction could at times increase the error. Instead, he offered a

[18] Recorde, *Whetstone of Witte*, sig. Hii^r.
[19] Recorde's defintion also called "surd nombers' "nombers irrationalle because many of thenn are soche, as can not bee expressed, by common nombers Abstracte, norther by an certain rationale nomber. Other men call them more aptly Surde nombers." Basically, he is stating that such numbers can not be expressed as integers or rational numbers.
[20] Recorde, *Whetstone of Witte*, sig. Lii^v.

better way to "gette verie nighe unto the roote. And you shall go as nighe as you will desire, in any practike worke."[21] Basically, one first decides how close an approximation is necessary. Then, one returns to the number for which one is extracting the root, adding a varying amount of zeros to the end of the number depending on the accuracy required. To get within 1/10th of the true root, Recorde advises his student to add two zeros; to get within 1/100, four zeros. In the above example, they decide to add six zeros to get within 1/1000 of the "true roote." At the end of the new computation of the root of 296,882,000,000, the student obtains 544,868, so the approximation achieved is 544 868/1000. It is very clear from Recorde's treatment of this example that he realises that one can arrive at closer and closer approximations through these methods.

Recorde claimed that the origin of his interest in these numerical methods resided in their practical uses. The Master states that the extraction of roots "serveth in so many waies, in building, . . for measuring ground Timber, or Stone. . . "[22] He then goes on to give specific useful examples, especially from "warre" such as scaling walls, engines and how many men might be needed in battalions. Recorde gave his student an example of a problem concerning a "capitaine general" who had three armies that he wished to form into square "battailes," but who did not know how many men to set in the front of each "battaile."[23] The master gives the scholar three numbers: the first army has 5625 men, the second has 9216 men, and the third 15129 men. The student replies that he must find the square roots of these numbers because that will give him the front, or flanke, of the battle. The student solves the problem by arguing that the three armies should have as front rows

i) $\sqrt{5625} = 75$
ii) $\sqrt{9216} = 96$
iii) $\sqrt{15129} = 123$

75 square = 5,625
96 square = 9, 216
123 square = 15, 129

Total *29,970.*

At this point, turning to the question of forming one large square battalion, the student points out that this number is not a perfect square, and he goes on to find the closest square number (29,929*)* and its root (173). Thus (173) "should bee the fronte of this greate battaile." This may seem to be an artificial problem — it is difficult to envision captains pausing before battles to compute square roots— but the task of finding square battalions seems to have been a popular one in this period. Leonard

[21] Recorde, *Whetstone of Witte,* sig. Lii^v.
[22] Recorde, *Whetstone of Witte*, sig. Mii^r.
[23] Recorde, *Whetstone in Witte*, sig. Mii^r.

and Thomas Digges, in *Stratioticos* (1579), also examined this question, as well as related problems dealing with square batteries of pike-men and the like.[24] *Stratioticos*, unlike Recorde's work, was very popular: it went through four editions (1572, 1579, 1585 and 1590). Recorde was obviously trying to offer examples of problems that would interest his audience, but given that it was never reprinted it seems safe to say that he failed, even though he provided many other examples of practical situations where one might need to find a square root.

Recorde proceeded to describe another kind of number, described as "*Cossike*, are such as bee contracte unto a denomination of some *Cossike* sign as 1. nomber.1.square.1.Cube."[25]

A list of cossic signs was provided:

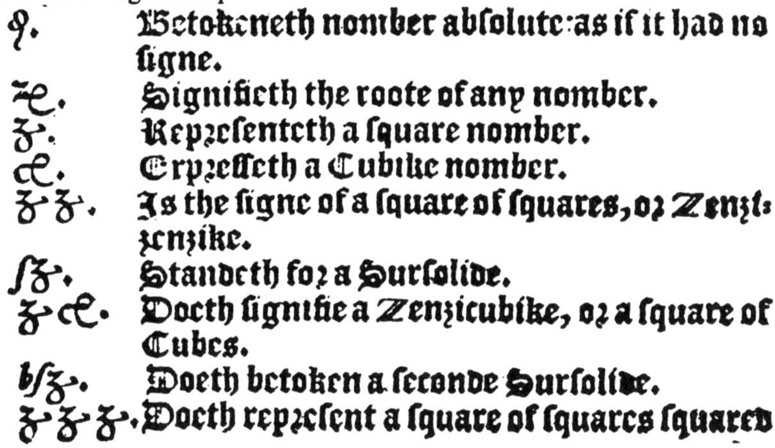

Figure 2[26]. *Cossic Symbols*

As the first line shows, there was a sign or symbol that signalled that a number was **not** associated with an unknown, which Recorde called "absolute." The next symbol, which Recorde associates with "the root of any number", was supposed to designate the unknown as long as it was not raised to any power. Different symbols were introduced to represent the *same* unknown when it was raised to various powers. He built up higher powers through a combination of these signs. Some of his descriptions can be misleading, in the light of his later usage, as when he labels one symbol the "roote of any nomber" when it is actually used as an unknown, or the "root" of the equation. Irrational numbers were in a way subsumed into the cossic numbers (numbers associated with variables or unknowns) as Recorde states: "And although there be many kindes of irrationall nombers, yet those figures that serve in Cossike nombers, bee the figures also of all irrational nombers. . . ."[27] In

[24] Leonard and Thomas Digges, *Stratioticos*, (London, 1579), p.52.
[25] Recorde, *Whetstone of Witte*, sig. Si^{v}-Sii^{r}.
[26] Recorde, *Whetstone of Witte*, sig. Si^{r}.
[27] Recorde, *Whetstone of Witte*, sig. Sii^{r}.

other words, cossic numbers could stand for irrational. Indeed, Recorde seemed to assume that in many cases it would be easier to express an irrational in cossic notation. Although Recorde is clearly concerned about the proper labelling and classification of irrational numbers, he is willing to work with them in practice. He went on to give examples of how multiplication, division, and fractions can be applied to cossic numbers (cossike was the English spelling).

Sometimes during his explanations we can catch a glimpse of Recorde's views about the nature of numbers. At one point the margin of the text declares "An Absurde nomber expresseth lesse than naught".[28] In other words, for Recorde, negative numbers were to be considered absurd. Nevertheless, he allowed negative cossic numbers to appear in his equations.

After making sure his student was confident operating with cossic numbers, Recorde introduced "The rule of equation, commonly called Algebers Rule":

> When any question is propounded
> apperteinyng to this rule, you
> shall imagin a name for the nom=
> ber, that is to bee soughte, as you
> remember, that you learned in
> the rule of false position. And with that nomber
> shall you procede, accordyng to the question, until
> you finde a Cossike nomber, equalle to that nom=
> ber, that the question expresseth, which you shal reduce
> ever more to the leaste nombers. And then
> divide the nomber of the lesser denomination, by
> the nomber of the greateste denominatio, doe beare the signe
> of some rooted no[m]ber. For then must you extract
> the roote of that quotiente, accordyng to
> that signe of the denomination.[29]

In other words, Recorde wanted the student first to assign the unknown a "name;" this would be the cossic symbol for the unknown. Then one would add, subtract, multiply or divide until the cossic number was equal to a number. In some cases, one would be left with a quadratic equation. After the statement of the above rule, the student first supposes that this rule is the same as the rule for false position, but the Master claims that "Algebers" rule is better because it uses "true" numbers instead of "false." He then goes on to explain equations using two cases. The first is when one number is compared as equal to another; the second, is when one cossic number is compared as equal to two other cossic numbers. It is at this point that Recorde introduces his '=' sign for equality "to avoid the tediouse repetition of the wordes: is equalle to."

Next, Recorde deals with "simple nombers Cossike, and uncompounded," such as

---

[28] Recorde, *Whetstone of Witte*, sig. Bbiii^v.

[29] Recorde, *Whetstone of Witte*, sig. Eeiv^v-Ffi^r. The quotation follows the original printer's layout of Recorde's text.

or in modern notation,

$$4x^2 = 12x. \qquad (1)$$

In all the examples dealing with this type of equation Recorde uses simple division to solve the problem. The subsequent case is where "one simple nomber Cossike, is compared as equalle to .2. other simple numbers Cossike". For example, for the equation

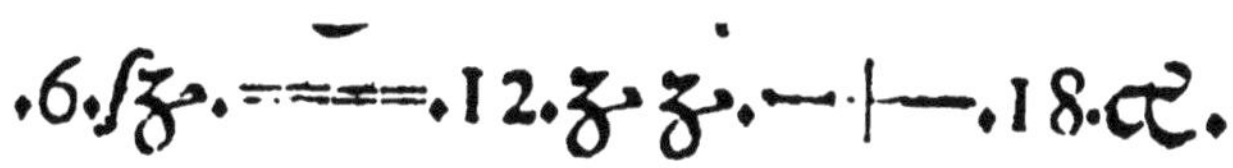

or in modern notation,

$$6x^5 = 12x^4 + 18x^3, \qquad (2)$$

by using division and perhaps inspection, the Master obtains the root 3, without explaining how he arrived at it. The other solution, -1, was ignored. Next, he quickly moved on to solving practical problems using these techniques. Examples dealt with armies, walls, bricks, wills, and general mercantile problems.

Some problems, although they are dressed up as applications, seem to stem more from mathematical curiosity. A captain has a great army, which he wants to form into a square battalion which is as large as possible. Apparently, the captain didn't mind stopping to perform some computations. The Master tells the student that on the captain's first try, he had 284 men left over outside the square. When he added "1 more in the front" (increased his guess of the size of the side of the square by 1), he had twenty-five men too few to form the square. The Master then asked the student how many total soldiers the captain had.

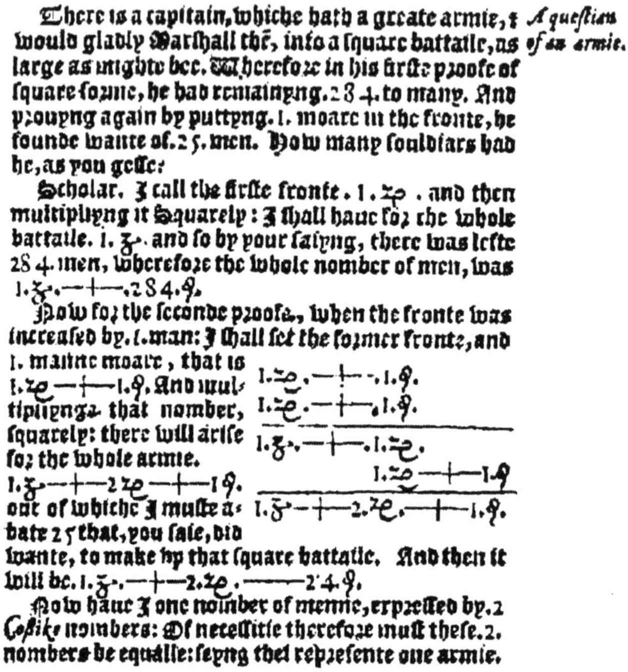

There is a capitain, whiche hath a greate armie, & would gladly Marshall thē, into a square battaile, as large as mighte bee. Wherefore in his firste proofe of square forme, he had remainyng. 284. to many. And prouyng again by puttyng. 1. moare in the fronte, he founde wante of. 25. men. How many souldiars had he, as you gesse? *A question of an armie.*

Scholar. I call the firste fronte. 1. ꝛ. and then multiplying it Squarely: I shall haue for the whole battaile. 1. ʒ. and so by your saiyng, there was lefte 284. men, wherefore the whole nomber of men, was 1. ʒ. —+— 284. φ.

Now for the seconde proofe, when the fronte was increased by. 1. man: I shall set the former fronte, and 1. manne moare, that is 1. ꝛ —+— 1. φ. And multipliyng that nomber, squarely: there will arise for the whole armie. 1. ʒ —+— 2 ꝛ —+— 1 φ. out of whiche I muste abate 25 that, you saie, did wante, to make vp that square battaile. And then it will be. 1. ʒ. —+— 2. ꝛ. —— 24. φ.

1. ꝛ. —+— 1. φ.
1. ꝛ. —+— 1. φ.
1. ʒ. —+— 1. ꝛ.
1. ꝛ —+— 1. φ
1. ʒ —+— 2. ꝛ. —+— 1. φ.

Now haue I one nomber of menne, expressed by. 2 Cossike nombers: Of necessitie therefore muste these. 2. nombers be equalle: seyng thei represente one armie.

Figure 3. A question of an armie.

The student solved the problem in this way: He set the first front (side of square) = x so the square battalion = x^2. Therefore on the captain's first try: The entire army was set equal to $x^2 + 284$. Then, on the captain's second try, he added 1 to the first front, or, first front was set equal to x + 1. Squaring this he obtains,

$$(x+1)(x+1)= x^2 + 2x +1, \tag{3}$$

so again, the entire army equals

$$x^2 + 2x +1 - 25 = x^2 + 2x -24. \tag{4}$$

Then student finds that:

$$x^2 + 2x -24 = x^2 + 284, \text{ so } x = 154. \tag{5}$$

This problem seems contrived. First, it seems unlikely that any military gentleman would really need to know anything like this except in the context of a game, or puzzle. Second, as with many of Recorde's questions, the problem is set up so that the solution is a positive whole number.

Recorde then went on to discuss irrational numbers, or as he calls them "surde" numbers. "Surde" numbers are defined as "nothing els, but soche a nomber set for a roote, as can not be expressed by any nomber absolute."[30] In other words, if one squares a "surd" number, one does not obtain an integer. A distinction, by examples, is made between such numbers whose roots were positive integers and therefore "noe Surde nomber", and numbers such as √5 which cannot be expressed as a positive integer. The "Surdness" of a number seems for the most part to have been decided by whether a number had its square, cube, or fourth root (which ever is required) expressible as a positive whole number. He seems content to express "surde" numbers symbolically without worrying about their foundational status. He gives no practical example to justify his discussion; "surde" numbers seem to occupy a more abstract position than "broken" or "absolute" numbers. The whole section describing irrationals is only about three pages long. Recorde mainly supplied examples of numbers that are "surde", and then he quickly moved on to how one can operate with these numbers, explaining addition, subtraction, division and multiplication.

In an example illustrating addition, he mentions commensurability. Recorde's goal is to discover whether when trying to add "surd" number, they could somehow be simplified to obtain one number under the radical sign. So, for example, when adding √384 + √150, 384 and 150 are commensurable, if they can be manipulated into one number under the radical sign (Recorde realised that this would not always be possible).

He defines commensurability in practice then as "soche nombers, as by one common divisor, maie be brought into Square nombers."[31] Using 384 and 150 as an example, the Master tells the student that the first step is to find (in modern terms) the greatest common divisor. Recorde does not explain that this division would leave you with two numbers that were relatively prime. In this case, one would divide both numbers by six and obtain 64 and 25. The check to ensure commensurability was to discover whether these numbers were perfect squares. If they are, one may proceed by a sort of unrolling and re-rolling process. Finding the squares, 8 and 5, one adds them. Then one take the square of the result of the addition, 13, obtaining 169, at this point, one multiplies the greatest common divisor back in (6 * 169 = 1014). Finally the answer to √384 +√150 = √1014. It seems likely that as approximating roots was a time consuming process, Recorde's goal was to reduce the amount of times this might be required. This definition and these procedures, however, did not lead to any kind of treatment of commensurability. Additionally, Recorde gave no reason why his procedure worked. Many examples are provided, so the student could copy the Master in a recipe-like way. None of his discussion of "surde" numbers makes any geometric references.

The *Whetstone* ends abruptly at this point with a knock on the door from a messenger who calls the Master away. Recorde promises to return at a later date to his discussion of "surde" numbers, but his early death (in King's Bench prison for

---

[30] Recorde, *Whetstone of Witte*, sig. Lliij^r.
[31] Recorde, *Whetstone*, sig. Nnii^v.

debt) prevented him from writing a sequel, and the *Whetstone of Witte* was his last book. The *Whetstone* is the only one of Recorde's books not published in several editions, and it is doubtful that the craftspeople, merchants, artisans, and military leaders for whom this work was said to be written needed these relatively advanced computational skills.[32] Even if Recorde had lived to write another book describing in more detail such complicated topics as "surde" numbers, it is unlikely that it would have been any more popular than the *Whetstone*.

In general, while Recorde maintained an Aristotelian concept of "absolute" numbers being discrete, in practice he was willing to extend numeric operations to "broken" and "surde" numbers. Klein's thesis seems substantiated to a certain extent by Recorde's greater acceptance of irrationals when they were linked to cossic numbers. Perhaps this association did indeed help numbers to appear to be more symbolic. Although numbers in Recorde's work were not interchanged with geometric magnitudes, this was partially due to the nature of the problems solved. He was willing to work with a broadened number concept, but he did not seem concerned with providing the new types of calculation with a rigorous foundation, being more interested in practical results. His interest in labelling non-traditional numbers and his mentioning of Aristotle's definition of numbers, however, indicate that there was some tension between the demands of practice and theoretical concerns. Practically speaking, positive numbers were in some ways treated as though they are continuous, in that Recorde operated on fractions and irrational numbers. For instance, he gave fractional expansions when obtaining the roots of numbers that were not perfect squares. He seems to have understood the value of approximating his results to the desired degree of accuracy.

Recorde's work documents how algebraic practice could lead to the abandonment of traditional Greek concepts of number. The demands of war, commerce, and measurement motivated a search for new numeric solutions that treated numbers in a new way, insofar as the manipulation of the equations could not always be carried out solely in terms of positive integers. Moreover, as Klein's theory suggests, Recorde also used irrationals and fractions more frequently when he utilised cossic symbolism. In Recorde's time, however, although there was an intense interest in practical applications of mathematics, buyers of mathematical texts were not as yet deeply interested in the algebraic techniques themselves.

THE TREATMENT OF NUMBERS IN ARITHMETIC TEXTS

Humfrey Baker's *The Well Springe of Sciences* (London, 1562), although it did not include cossic notation or equations, was similar in certain respects to Recorde's *Whetstone of Witte*. The text was dedicated to the Merchant Adventurer's Company, and it was also devoted to practical aims. Interestingly, Baker also adhered to the traditional notion of number; number was again defined in an Aristotelian fashion as

---

[32]Easton, *Dictionary of Scientific Biography*, p.339.

"a multitude composed of many unities . . . And therefore an unity is no number, but the beginninge and origin of number."[33] It is clear that Baker was familiar with, and accepted, the traditional Euclidean/Aristotelian number concept. As with Recorde, however, practical necessities created tensions in Baker's depiction of numbers. A broadened number concept was needed to provide solutions for a variety of problems. Therefore, fractions, or "broken numbers", were defined as "a parte or manye partes of 1".[34] Fractions were mainly used in problems dealing with money, weight and measure. Baker was particularly concerned with conversion of measurements and monetary units from different cities such as Paris, Lyons, and Vienna.

In general, Baker's work seems better suited towards merchants' needs than Recorde's. His theoretical discussions were kept to a minimum, and he concentrated on solving abacus style problems without introducing algebraic complications. Although he mentioned the traditional definition of number, he quite pragmatically utilised the idea of fractions in his sections dealing with questions of "losse & gain, Tapostrie [tapestry], Fellowship [assigning relative gain in partnership situation], and False position." Keeping in mind the lesser scope of Baker's text, he seems to share Recorde's willingness to provide a traditional definition of number while simultaneously utilising a slightly broader number concept in practice.

Blundeville's *Exercises* (1594) were wider in scope, include arithmetic, trigonometric tables, the use of Globes, description of other instruments, and a section on mathematical navigation. His arithmetic was intended to be practical, and "so plainlie written as any man of a mean capacitie may easilie learn the same without the helpe of any teacher."[35] Since it was only a small section of the text, however, few examples were provided.

Blundeville defines number fairly traditionally as "a collection or somme of many ones added together."[36] Fractions were said to be the "broken partes of some whole thing . . . ", and compared to smaller coins that together equalled the amount of a greater coin. Square roots are also discussed, but no mention is made of what one should do in cases where one is not taking the root of a perfect square, and there is no mention of surd, or irrational, numbers. Once again, square roots seem to be of interest mainly for the "setting of battels", and the examples all take the root of perfect squares. Because of the amount of material covered, and the rushed nature of the presentation, it is difficult to tell why the text leaves out complications. It could simply be that as one would not usually break up men into fractional sections to form battalions, it was not considered to be necessary.

Blundeville's work, like Baker's and Recorde's, shows a willingness to deal with numbers that were not merely positive integers. Fractions, in particular, served many practical uses in the sections of the text that deal with maps and navigation. Although his limited treatment of roots and his neglect of cossic material makes

---

[33] Humfrey Baker, *The Well Springe of Sciences* (London, 1562); I'm quoting from an augmented 1574 edition, Sig. B.i.^{r}.

[34] Baker, *The Well Springe of Sciences*, fol. 50.

[35] Blundeville, *His Exercises Containing Sixe Treatises*, (London, 1594), Sig. A3^l.

[36] Blundeville, *His Exercises Containing Sixe Treatises*, p. 1.

direct comparisons with Recorde's work difficult, it seems that he shared in the general willingness to move slightly beyond positive integers.

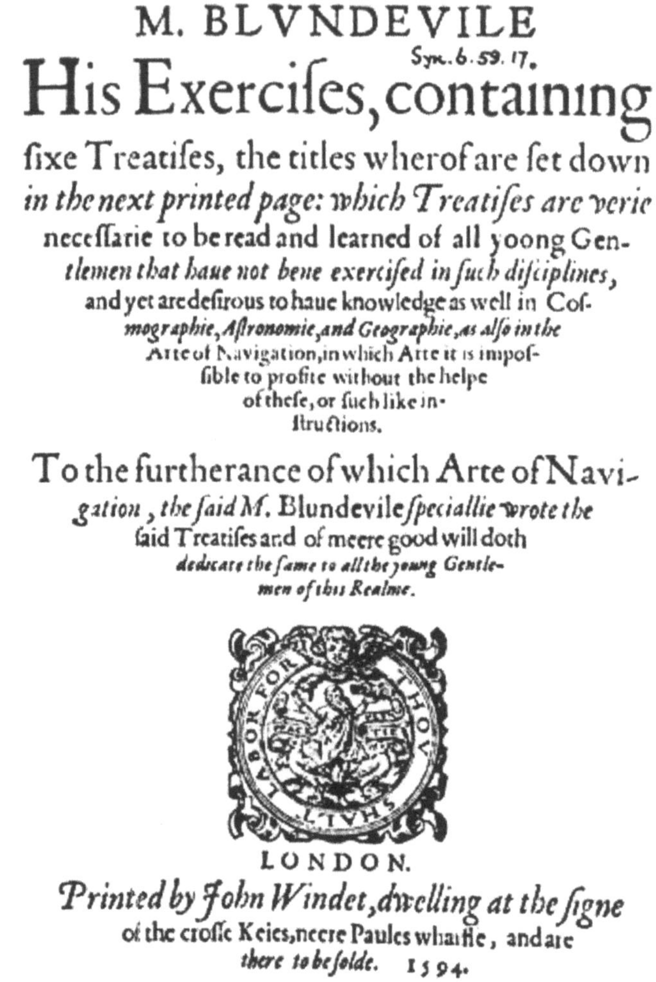

M. BLVNDEVILE

Syn.6.59.17.

His Exerciſes, containing ſixe Treatiſes, the titles wherof are ſet down *in the next printed page: which Treatiſes are verie* neceſſarie to be read and learned of all yoong Gen- *tlemen that haue not bene exerciſed in ſuch diſciplines,* and yet are deſirous to haue knowledge as well in Coſ- *mographie, Aſtronomie, and Geographie, as alſo in the* Arte of Nauigation, in which Arte it is impoſſible to profite without the helpe of theſe, or ſuch like inſtructions.

To the furtherance of which Arte of Naui- *gation, the ſaid M.* Blundevile *ſpeciallie wrote the* ſaid Treatiſes and of meere good will doth *dedicate the ſame to all the yoong Gentle- men of this Realme.*

LONDON.

Printed by John Windet, dwelling at the ſigne of the croſſe Keies, neere Paules wharffe, and are *there to be ſolde.* 1594.

Figure 4. Blundeville's Exercises

The treatment of numbers in early modern English arithmetics seems to demonstrate the importance of algebra to the broadening of number concepts. Many of the specific problems solved were very similar to those proposed by Recorde. Although the arithmetics made use of fractions, and seem willing to accept in practice "numbers" that were not collections of units, they did not stretch the idea of number to include surds or indefinite approximations by fractions. An interest in solving practical problems was not, in itself, enough to cause a broadening of number concepts. The solutions of many of the story-problems dealt with in the arithmetics could, for all practical purposes, easily be approximated in fractional form. In many cases it is unclear how much exactness was required. More importantly, certain mathematical methods, coupled with an increased use of symbolism, also seem to be necessary. Without symbols that could stand interchangeably for numbers of many different types, or approximation techniques that went beyond simple fractions, number could not undergo more than a minor transformation.

JOHN TAPP

Tapp's *The Path-Way to Knowledge* (1613) once again included cossic material. In many ways Tapp's book is similar to Recorde's *Whetstone of Witte*. Both works were written in the form of a dialogue between a teacher and a student, were informal, and concentrated on practical examples. Tapp's text is (like Baker's and Blundeville's) if anything even less theoretically oriented than Recorde's. Hindu-Arabic numerals and place notation are described, but there is no general discussion concerning the nature of numbers. Fractions are described as "broken numbers", or as a "part of another number", and are operated on unproblematically. Many monetary examples are given; like Baker, Tapp also is especially interested in currency conversion.

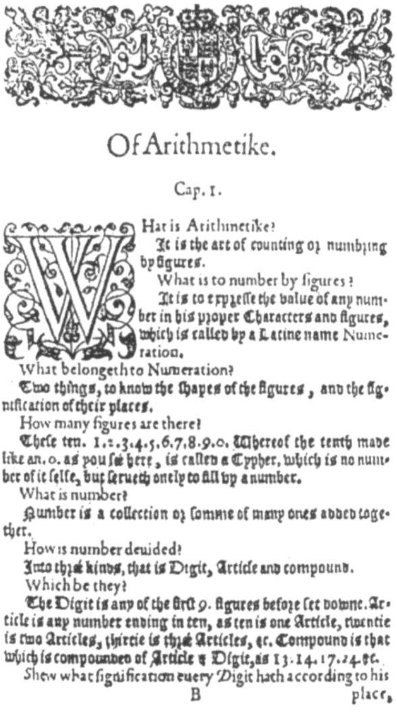
Of Arithmetike.

Cap. I.

WHat is Arithmetike?
It is the art of counting or numbring by figures.
What is to number by figures?
It is to expresse the value of any number in his proper Characters and figures, which is called by a Latine name Numeration.
What belongeth to Numeration?
Two things, to know the shapes of the figures, and the signification of their places.
How many figures are there?
These ten. 1.2.3.4.5.6.7.8.9.0. Whereof the tenth made like an. o. as you see here, is called a Cypher, which is no number of it selfe, but serueth onely to fill up a number.
What is number?
Number is a collection or somme of many ones added together.
How is number deuided?
Into three kinds, that is Digit, Article and compound.
Which be they?
The Digit is any of the first 9. figures before set downe. Article is any number ending in ten, as ten is one Article, twentie is two Articles, thirtie is three Articles, &c. Compound is that which is compounded of Article & Digit, as 13.14.17.24.&c.
Shew what signification euery Digit hath according to his
B place,

Figure 5. Tap's definition of number

The extraction of roots and cossic numbers are linked for Tapp, as they had been for Recorde. Additionally, the finding of roots, for Tapp, is once again motivated by a wide range of practical applications; he claims that it is a skill useful for finding distances, heights, measuring land, navigation, and, of course, the setting of battles. His text is closer to Recorde's than to Baker's or Blundeville's in recognising complications. While discussing the extraction of roots, Tapp stated,

> The difficulty in working . . .if it be not a right square number, then the true roote therof cannot be exactly found, and therefore your best indeauour must be to find the nearest roote, and that may have the least error.[37]

Tapp then goes on to show how, in some cases, one can find closer and closer approximations, although never an exact root. It is implied, although not stated explicitly, that this process could go on indefinitely.

When Tapp moved to his explanation of "the Art of Cossicke numbers", he began with an explanation of the symbols (which were the same as Recorde's, except that he used **N** to represent a number not attached to an unknown). Next, he moved through the addition, subtraction, multiplication, and division of cossic numbers. Finally, he separated equations into four types in a manner similar to Recorde. Equations were solved by simplification and division. There was no awareness, however, of the possibility of surd solutions or the need for approximation techniques.

Although the "art of Cossicke numbers" could, as in Recorde's case, lead to a broadening of the number concept, we see in Tapp's case that this was not inevitably the case. Tapp avoided, as had the earlier arithmetics, any expansion beyond fractions, although he was aware that in taking some roots one might need a more complex approximation to find a more exact solution. Cossic notation could, in the right hands, allow for numbers to be associated with irrationals, or with magnitude, but not every mathematician took that step. In Tapp's work we can see that it is possible to avoid techniques and situations that lead to a stretching of the number concept. It was not until the introduction, by Harriot and Oughtred, of a more "modern" style of algebra into England, that number concepts were considered again.

THOMAS HARRIOT'S *ARTIS ANALYTICAE PRAXIS*

After receiving his Bachelor of Arts degree from Oxford, around 1583 Harriot began his association with Sir Walter Ralegh, who employed Harriot as an expert in cartography and mathematical navigation. By the 1590s, however, Harriot had turned to Northumberland for patronage. Although Harriot was a prolific writer with wide-ranging natural philosophical interests in optics, astronomy, and ballistics, Harriot in his lifetime published only one book, *A Briefe and True Report on the New Founf Land of Virginia.*[38] His mathematical work, the *Praxis*, was published

---

[37] Tapp, *The Path-Way to Knovvledge,* p. 302.

[38] London (1588)

posthumously from material gathered from his voluminous manuscript notes. Harriot's *Artis Analyticae Praxis*, published in 1631, differs from the works of Recorde and Oughtred in that it is a sample of his papers in algebra published ten years after his death by Walter Warner.[39] It includes only the particular algebraic problems that Warner selected from Harriot's mathematical manuscripts, and it therefore provides only a limited understanding of the scope of Harriot's work. Warner was also a mathematician and physicist and Harriot shared with Warner the patronage of Henry Percy, the Ninth Earl of Northumberland, known as the "Wizard Earl".[40]

Harriot did not choose the material himself, and the text was intended to be as much a testimony of Harriot's achievements as an algebra textbook. The *Praxis* is also unusual in that little that is now recognised as Harriot's pioneering work was selected for publication.[41] Although the *Praxis* does not contain the best of Harriot's mathematics, it was still influential because John Wallis included a summary of it (which perhaps gives Harriot credit for more mathematical advances than he deserves) in his *Treatise of Algebra*. Indeed, the mythology surrounding Harriot's achievements might have been more influential than his published text. This section will first discuss Harriot's reputation and place in the English mathematical community and then will investigate his *Praxis* and Wallis' summary of its contents. Although the *Praxis* does not contain Harriot's finest mathematical work, it was the only part of his technical writing that was readily available. Later mathematicians, such as Wallis, were unable to find Harriot's manuscripts. Secondly, an overview of Harriot's views on mathematical continua will be presented. It is difficult to be certain, however, of how generally influential these views were outside Harriot's own circle, but Harriot's notions of continua at least shows that it was possible in this period for a English or Scottish mathematician to have discarded the traditional Aristotelian view.

Harriot enjoyed considerable contemporary fame both as an explorer and as a mathematician.[42] Oughtred's *Clavis*, which was published in the same year as Harriot's work, was inferior to the *Praxis* in originality, scope and suggestiveness as a textbook. The *Clavis,* however, had the advantage of being considered excellent, small and available.[43] Therefore the *Clavis* had more direct influence on the English teaching of algebra than did the *Praxis*. The *Praxis*, however, was not unknown. As late as 1630, Hakewill mentioned Harriot's work in his *Apology* as an example of

---

[39] Richard W. Hadden, *The Social Origins of Early Modern Science* (McMaster University: 1984), p.205.

[40] Robert Kargon, *Atomism in England from Harriot to Newton* (Oxford: Clarendon Press, 1966), p.6. The Ninth Earl of Northumberland (1564-1632) is now remembered for his patronage of scholars and scientists including Thomas Harriot, Walter Warner and Robert Hues. In his own time he was widely admired for his studies in natural philosophy.

[41] Edward Rosen, "Harriot's Science: The Intellectual Background," in *Thomas Harriot: Renaissance Scientist*, ed. John Shirley (Oxford: Clarendon Press, 1974), p.103.

[42] John Shirley, "The Scientific Experiments of Sir Walter Raleigh, the Wizard Earl, and the Three Magi in the Tower, 1603-1617," in *A Source Book for the Study of Thomas Harriot*, ed. by John Shirley (New York: Arno Press, 1981), p.58.

[43] Frank Morley, "Thomas Harriot: 1560-1621," in *A Source Book for the Study of Thomas Harriot*, ed. by John Shirley (New York: Arno Press, 1981), p.63.

modern improvements on ancient mathematics.[44] Wallis also highly praised Harriot's work on algebra.[45] Additionally, Wallis was the source of another widespread rumour: that Descartes plagiarised Harriot. It is probable that Wallis was too exuberant in his praise of Harriot.

Viète's algebra had come out from 1591 through 1600, and it may be supposed that Harriot's work was contemporary. Yet, as Sir William Lower stated, "[Harriot's] great reservednesse" and "much procrastination" has hindered us from knowing the full extent of his work.[46] Florian Cajori suggested that Harriot learned from Viète to use letters to represent numbers in equations, using vowels for unknowns and consonants for known numbers, although unlike Viète, he employed small letters instead of capitals.[47] The contents of the *Praxis* did follow Viète, although there were improvements in notation and some simplification in technique. One of the chief innovations in the book was bringing together all the terms of an equation and equating them to zero. One must keep in mind when reading the *Praxis* that Harriot's standpoint was primarily that of a physicist. Harriot left over 10,000 folio pages of scientific papers that discuss experimental and theoretical work in several fields and these documents abound with measurements, diagrams, tables, and calculations.[48] Harriot's interest in numerical solutions was shaped by his concern for physical phenomena; negative roots do not help you to solve problems in free fall and in fall in resistant media.

In no other book of this period do formulas stand out so prominently.[49] The only real text appears in the preface and in the definitions. After that, words are practically dispensed with. Harriot quickly demonstrated how to add, subtract, multiply and divide with symbols within the space of two pages.[50]

---

[44] J.A. van Lohne, "Thomas Harriot als Mathematiker," *Centaurus* 11 (1965), p. 31.

[45] Jean Jacuot, "Thomas Harriot's Reputation for Impiety," in *A Source Book for the Study of Thomas Harriot*, ed. John Shirley (New York: Arno Press, 1981), p.165.

[46] Frank Morley, "Thomas Harriot: 1560-1621," in *A Source Book for the Study of Thomas Harriot*, ed. by John Shirley (New York: Arno Press, 1981), p.64.

[47] Florian Cajori, "A Revaluation of Harriot's *Artis Analyticae Praxis*," *Isis* 11 (1928), p.316.

[48] J.A. Lohne, *DBS*, p. 124.

[49] Florian Cajori, "A Revaluation of Harriot's *Artis Analyticae Praxis*," p. 318, agrees with this point although in general he found the *Praxis* to be overrated.

[50] Thomas Harriot, *Artis Analyticae Praxis* (London: 1631), p.7-9.

4 DEFINITIONES.

muniter deducitur æquationem communem siue aduenticiam nuncupare licet, vt ipso nomine ab alio æquationis genere prorsus diuerso distinguatur, quam canonicam appellare visum est, de qua postea.

DEFINITIO 14.

Est etiam aliud quoddam æquationum genus; quæ licet canonicæ non sint, cum tamen ab ijs, vt ab originalibus suis æquationes canonicæ deriuentur, canonicarum originales in sequentibus denominantur. Huius generis æquationes per genesim siue multiplicationem e radicibus binomiis absque alio discursu immediatè conficiuntur. In quibus facta ex radicibus multiplicatis, radicibus ipsis multiplicandis sub forma tantùm multiplicationis ordinatis æquantur, vt in exemplis hisce videre est

$$\begin{array}{l} a+b \\ a-c \end{array} = \begin{array}{l} aa+ba \\ \quad -ca-bc \end{array}$$

$$\begin{array}{l} a+b \\ a+c \\ a-d \end{array} = \begin{array}{l} aaa+baa+bca \\ \quad +caa-bda \\ \quad -daa-cda-bcd \end{array}$$

$$\begin{array}{l} a+b \\ a+c \\ a+d \\ a-f \end{array} = \begin{array}{l} aaaa+baaa+bcaa \\ \quad +caaa+bdaa \\ \quad +daaa+cdaa+bcda \\ \quad -faaa-bfaa-bcfa \\ \qquad -cfaa-bdfa \\ \qquad -dfaa-cdfa-bcdf \end{array}$$

Harum æquationum forma propriæ æquationis definitioni minimè quadrat. Quæ tamen ab ijs deriuantur duæ canonicarum species, primaria scilicet & secundaria, tum definitioni conformes sunt, tum vsui propriè applicabiles.

DEFINITIO 15.

Primaria canonicarum species est earum quæ ab originalibus per deriuationem constituuntur. Nam reiectâ formali radicum ordinatione, quæ æquationis originalis pars altera est, & ex alterius partis homogeneis homogeneo dato per affectionis mutationem in situm reliquis oppositum translato, fit æquatio canonica primaria. Deriuationis ratio in sectione 2. proponitur, speciei autem exempla hæc sunt.

$$\begin{array}{l} aa+ba \\ \quad -ca = +bc \end{array}$$

$$\begin{array}{l} aaa+baa+bca \\ \quad +caa-bda \\ \quad -daa-cda = +bcd \end{array}$$

$$\begin{array}{l} aaaa+baaa+bcaa \\ \quad +caaa+bdaa \\ \quad +daaa+cdaa+bcda \\ \quad +faaa-bfaa-bcfa \\ \qquad -cfaa-bdfa \\ \qquad -dfaa-cdfa = +bcdf \end{array}$$

DE-

Figure 6 (continues next page). Harriot's Equations

12

SECTIO SECVNDA.

Æquationum Canonicarum ab originalibus suis derivatio *sive deductio:*

Præmissâ ipsarum originalium è radicibus binomijs per genesin constitutarum ordinatâ hac descriptione.

Quadraticæ.

1. . . . $\begin{array}{l} a-b \\ a-c \end{array} \Big| = \begin{array}{l} aa-ba \\ \quad -ca+bc \end{array}$

2. . . . $\begin{array}{l} a-b \\ a+c \end{array} \Big| = \begin{array}{l} aa-ba \\ \quad +ca-bc \end{array}$

3. . . . $\begin{array}{l} a+b \\ a+c \end{array} \Big| = \begin{array}{l} aa+ba \\ \quad +ca+bc \end{array}$

Cubicæ.

1. . . . $\begin{array}{l} a-b \\ a-c \\ a-d \end{array} \Big| = \begin{array}{l} aaa-baa+bca \\ \quad -caa+bda \\ \quad -daa+cda-bcd \end{array}$

2. . . . $\begin{array}{l} a-b \\ a-c \\ a+d \end{array} \Big| = \begin{array}{l} aaa-baa+bca \\ \quad -caa-bda \\ \quad +daa-cda+bcd \end{array}$

3. . . . $\begin{array}{l} a+b \\ a+c \\ a-d \end{array} \Big| = \begin{array}{l} aaa+baa+bca \\ \quad +caa-bda \\ \quad -daa-cda-bcd \end{array}$

4. . . . $\begin{array}{l} a+b \\ a+c \\ a+d \end{array} \Big| = \begin{array}{l} aaa+baa+bca \\ \quad +caa+bda \\ \quad +daa+cda+bcd \end{array}$

Reciprocæ Cubicarum.

5. . . . $\begin{array}{l} aa-bc \\ a-d \end{array} \Big| = aaa-daa-bca+bcd.$

6. . . . $\begin{array}{l} aa+bc \\ a-d \end{array} \Big| = aaa \quad\quad a-bcd$

7 aa—

Figure 6 (continued). Harriot's Equations

He next introduced the sign for equality, and his personal innovation, the signs for inequalities that we still use today.[51] His next step was to write down the expanded products of $(a - b)\ (a - c)$, $(a - b)\ (a + c)$, and $(a + b)\ (a + c)$ in order to set up three quadratic equations leading to what he called "canonical equations", that is, as products of binomial factors.[52] For example,

$$(a - b)(a + c) = aa - ba + ca - ba, \tag{6}$$

which results in an equation equivalent to what we would write in modern notation as

$$x^2 + (-b + c)x - bc = 0. \tag{7}$$

To solve this he places a = b, and writes $a - b = 0$, and multiplies both sides by $a + c$ to get

$$aa - ba + ca = bc \tag{8}$$

and b is given as the root. In the *Praxis* Harriot does not recognise $-c$ as a root, he does not place $a = -c$ or write $a + c = 0$. In his unpublished manuscripts, however, he gives the full solution of a quartic equation with the roots $+5, -7, 1 + \sqrt{-32}, 1 - \sqrt{-32}$ "written exactly as we write them."[53] So the recognition of only positive roots in the *Praxis* can be viewed as the choice of Warner. Although Rafael Bombelli and Girolamo Cardano recognised "impossible" solutions in the sixteenth century, negative and complex numbers were not regarded as having the same status as the positive integers, but were generally taken as symbolising the absence of a solution.[54] Warner might have felt it to be prudent to ignore them.

Harriot went on to perform thirty-two multiplications of binomial factors. For example he performed multiplications of the type

$$(a - b)(a - c)(a + d) \tag{9}$$

to get cubic polynomials. Once again he put $a = b$ or $a = c$ and ignored the root $-d$, although he was actually utilising factoring to find the roots.[55] His method was to show how certain canonical equations were actually built up from a multiplication of factors, making it simple to find the roots.

---

[51] Harriot, p.10.
[52] Harriot, p. 12.
[53] R.C.H. Tanner, "Henry Stevens and the Associates of Thomas Harriot," in *Thomas Harriot: Renaissance Scientist* (Oxford: Clarendon Press, 1974), p.101.
[54] Albert C. Lewis "Complex Numbers and Vector Algebra," in *Companion Encyclopedia of the History and Philosophy of the Mathematical Sciences*, ed. I. Grattan-Guiness (New York: Routledge, 1994), p. 723.
[55] Harriot, p. 17.

Wallis gives a detailed description with examples of this portion of the *Praxis*. He correctly notes the advances Harriot made in comparison with Oughtred and Viète in terms of notation, and mentions the improvement of putting the terms of the equation to one side and making the whole equal to zero.[56] Moreover, Wallis gives an excellent account of Harriot's use of canonical equations as paradigms or patterns for finding the roots of common equations. But he mistakenly attributes negative roots to the *Praxis*.[57]

The *Praxis*, however, either overlooks negative roots or rejects those equations that have only negative roots. For example, when discussing equations constructed out of factors such as $(a + b)(a + c)$, the *Praxis* states "since they cannot be derived except on the supposition of negative roots, they will be disregarded as quite useless."

Although Harriot did work with negative and complex roots in his manuscripts, there is no sign in the *Praxis* that he realised that every algebraic equation of nth degree has n roots. Yet in his manuscripts it is evident that he viewed the minus sign as a symbol in its own right and was comfortable with utilising negative numbers. For example he took the equation

$$x^2 - 6x + 25 = 0, \tag{10}$$

written as

$$25 = 6a - aa, \tag{11}$$

and rewrote it as

$$aa - 6a = -25. \tag{12}$$

He added 9 to both sides, whence

$$aa - 6a + 9 = -16 \text{ or } a - 3 = \sqrt{-16} \tag{13}$$

and

$$3 - a = \sqrt{-16} \tag{14}$$

leading to the two roots

$$a = 3 + \sqrt{-16} \text{ and } a = 3 - \sqrt{-16}\,.^{58} \tag{15}$$

---

[56] John Wallis, *A Treatise of Algebra, Both Historical and Practical* (London: 1683), p.128.
[57] Wallis, *Algebra*, p.132.
[58] R.C.H. Tanner, "The Ordered Regiment of the Minus Sign: Off-beat Mathematics in Harriot's Manuscripts," *Annals of Science* 37 (1980), p.137.

Perhaps Wallis was confused by the rumours of the contents of Harriot's manuscripts into making claims for Harriot that went beyond the content of the *Praxis.*[59] We know that he did not actually read the manuscripts themselves as he wrote just before the publication of his *Algebra* "I have never read any of those things, but only that of his Algebra."[60]

Throughout the first portion of the *Praxis* it is difficult to ascertain Harriot's own views about fractions and irrational numbers. The difficulty is compounded by the fact that Warner seems to have included only a limited amount of the available material. Unfortunately, all of the examples given have integer roots, and usually only one positive root is offered. However, it should be kept in mind that Harriot did not pick out the examples himself. Moreover, his manuscripts show that he was familiar with negative roots. The choice of problems probably tells us more of Warner's mathematical beliefs than Harriot's. Although Warner was a physical atomist, he believed that space was "continuel, eternal, infinite, unalterable, immoveable, [and] absolutely simple."[61] Perhaps he did not approve of all of Harriot's mathematical work. Luckily, some of Harriot's manuscripts provide us with more details concerning his views on continua.

Harriot's papers contain a work titled *De infinitis* that did deal with continua. He had encountered the problem of the continuum through working on a problem of marine cartography: the improvement of the "plane charts" used for navigation at that time.[62] He started with Aristotle's *Physics*, Book VI, but he used notions of maximum and minimum introduced into the debate by the scholastics. He considered several questions: is the infinite composed of finite parts? Further, is the finite composed of indivisible parts and can it be resolved into indivisibles? Is there a transition from the finite to the infinite through a maximum finite? Like Galileo he also wondered if equality and inequality could be attributed to infinities and indivisibles. By means of geometrical constructions and the comparison of progression he seemed to think that notions of indivisibles and the universally infinite could be applied to the physical world.[63] But he was unable to prove that there is a limit of the divisibility of space. Moreover the manuscript gave no indication of how to solve the problems posed by many of his questions, and Harriot wrote ambiguously in English at the end of it, "Much ado, about nothing. Great warres and no blowes. Who is the foole now?"[64]

In the end, Harriot decided that a continuum is an aggregate of things in contact, or that continuity is composed primarily of indivisibles in contact with each other. The general tenor of his ideas was towards an atomistic conception of matter.[65]

---

[59] J.F. Scott, "John Wallis as a Historian of Mathematics," *Annals of Science* 3 (1936), p. 346.

[60] Scott, "John Wallis as a Historian of Mathematics," p. 347.

[61] Kargon, p.38.

[62] Amir Alexander, "The Imperialist Space of Elizabethan Mathematics", *Studies in History and Philosophy of Science* **26** (1995):559-592.

[63] B.L. Add. MS. 6782, ff. 362-364v.

[64] Hilary Gatti,*The Renaissance Drama of Knowledge: Giordano Bruno in England* (London: Routledge, 1989), p.65.

[65] John Henry, "Thomas Harriot and Atomism: A Reappraisal", *History of Science* **20** (1982): 267-296 points out that it is more doubtful that Harriot was actually a *mathematical* atomists.

Except for Nathaniel Torporley, all of the natural philosophers surrounding the Wizard Earl were atomists, as well as being Copernicans.[66] Harriot's atoms were eternal, hard and unalterable and empty spaces existed between them.[67] These concepts of atomism had been brought to England by Giordano Bruno between 1583 and 1585. Bruno was a major influence on the circle surrounding the Earl of Northumberland. The Earl owned six of Bruno's works, and in one of Harriot's unpublished papers there is a reference to Bruno's *De innumerabilibus, immenso, et infigurabili, seu de universo et mundis*.[68] It is possible that Bruno influenced Harriot to use Democritean principles.[69] Harriot imitated Bruno in addressing the questions about continua and infinities posed by Aristotle.[70] He may have been impressed by Bruno's declaration that "The basis for all error in physics and mathematics is the infinite division of the continuum."[71] In other words, he believed the problem to be ignorance of the minimum, or a refusal to acknowledge it.[72]

Although Harriot investigated mathematical indivisibles, he was more concerned with the properties of material bodies. The physical qualities of matter resulted from their magnitude, shape, and the motion of their atoms.[73] In several ways Harriot's was a *mechanical* philosophy. He used the growing importance of mining and the use of other machinery as a convenient physical analogy for natural philosophy. Harriot and his circle also relied solely upon matter and motion for their physical explanations. His physical atomism might have influenced him to adopt mathematical indivisibles, but Harriot was reluctant to make known his views on atoms and the void.[74] This is perhaps due to a long series of religious and political involvements with the English authorities. In the 1590s Harriot and another of his patrons, Sir Walter Raleigh, were under suspicion of atheism. Moreover, in 1591 Robert Parsons denounced Harriot as an Epicurean atheist and a conjuror.[75] But although he did not publish his ideas he was not an isolated figure. Harriot's manuscripts were disseminated through his disciples, but his only published

---

[66] Kargon, p.7.

[67] Jean Jacuot, "Thomas Harriot's Reputation for Impiety," p.180.

[68] Gatti, p.52. Nicholas Hill and Warner also make direct references to Bruno's work.

[69] Jean Jacquot, "Harriot, Hill, Warner and the New Philosophy," in *Thomas Harriot: Renaissance Scientist*, ed. by John Shirley (Oxford: Clarendon Press, 1974), p. 108.

[70] Gatti, p.58. Henry, however, severely criticised the claim of a Harriot/Bruno connection.

[71] Kargon, p.10.

[72] Alexander posits that Harriot was also influenced by the patterns of the new world discovery narrative. He believes that Harriot's work was very much shaped by the "Elizabethan expansionist enterprise." See especially pages 35 and 36.

[73] Kargon, p.26.

[74] The atomic theory was not controversial in itself, but several ideas were associated with it that were not in agreement with later Christian theology. For example, the idea that *ex nihilo nihi fit*, and that therefore the universe had always existed, was clearly in contradiction with Genesis. Moreover, the atomists were also associated with the idea that the soul is a material, mortal substance that is dispersed with the body after death. See David A. Hedrich Hirsch, "Donne's Atomies and Anatomies: Deconstructed Bodies and the Resurrection of Atomic Theory," *SEL* 31 (1991), pp. 71-72 and G. B. Stones, "The Atomic View of Matter in the Xvth, XVIth, and XVIIth Centuries," *Isis* 10, 1 (1928): 445-65.

[75] Kargon, p.27.

mathematical work remained his *Praxis*. In the *Praxis*, however, due to Warner's choice of material, Harriot's belief concerning mathematical continua are lost from view.

In some ways, what we have just described are the two faces of Thomas Harriot. The first face is the Harriot that was edited by Warner, the Harriot of the *Praxis*. This view of Harriot shows us a figure who made several improvements to algebraic practice such as his version of Viète's notation and his placing all the terms of the equation on one side and making the whole equation equal to zero. However his interesting ideas about mathematical continua and his acceptance of complex numbers are lost from sight. Wallis, for whatever reasons, tried to correct this account by attributing to Harriot's *Praxis* advances that cannot be found in the text. It is not, however, until we consider Harriot's manuscripts that we find material that is outstandingly original for England in this period. The second face of Harriot is that of a secret atomist. He believed that notions of indivisibles and physical atoms could be used as a tool for understanding the physical world. These beliefs in this period were grounds for religious accusations. Thus the whole circle of atomists surrounding the Earl of Northumberland did not make their ideas public and when atomism became acceptable in England this early groundwork was lost. Harriot's manuscripts, however, show that even in this early period it was possible to hold views concerning mathematical continua that were widely divergent from the traditional views of Aristotle.

OUGHTRED'S *CLAVIS MATHEMATICAE*

William Oughtred's (1574-1660) *Clavis Mathematicae* was one of the most influential mathematical texts in England from its publication in 1631 until John Wallis began to publish his research at Oxford in 1657. Although Oughtred was the rector of Albury, and never a professional mathematician, it was said "that he was a pittiful Preacher; the reason was because he never studyed it, but bent all his thoughts on the Mathematiques."[76] As a student at King's College, Cambridge, he mastered mathematics in addition to completing the normal course of study.[77] Soon, he was tutoring students in mathematics at both his own and other colleges.[78] Throughout his life he had many young men come into his home to learn

[76] John Aubrey, *Aubrey's Brief Lives*, ed. Andrew Clark (Oxford: Claredon Press, 1898), p.292.

[77] Although Feingold argues in *The Mathematicians' Apprenticeship: Science, Universities and Society in England, 1560-1640* (Cambridge: Cambridge University Press, 1984) that there was more mathematics taught in universities than is usually admitted, Oughtred claimed that he taught himself. In his "To the English Gentrie. . .the Just Apologie of William Oughtred" (1634), he stated "in Cambridge in Kings College....over and above the usuall studies I employed upon the Mathematicall sciences, I redeemed night by night from my natural sleep, defrauding my body, and inuring it to watching, cold, and labour, while most others tooke thier rest." fol. B^r.

[78] Florian Cajori, *William Oughtred: A Great Seventeenth-Century Teacher of Mathematics* (Chicago: Open Court Publishing Co., 1916), p. 4.

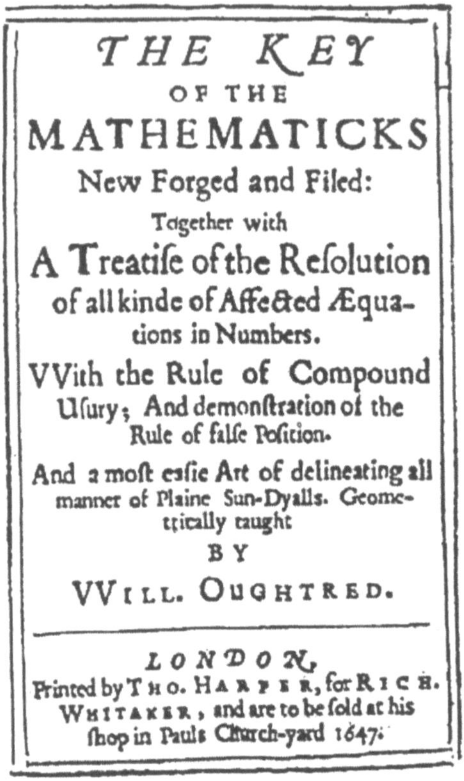
THE KEY
OF THE
MATHEMATICKS
New Forged and Filed:
Together with
A Treatiſe of the Reſolution of all kinde of Affected Æquations in Numbers.
VVith the Rule of Compound Uſury; And demonſtration of the Rule of falſe Poſition.
And a moſt eaſie Art of delineating all manner of Plaine Sun-Dyalls. Geometrically taught
BY
VVILL. OUGHTRED.

LONDON,
Printed by THO. HARPER, for RICH. WHITAKER, and are to be ſold at his ſhop in Pauls Church-yard 1647.

Figure 7. Oughtred's Key

mathematics. His *Clavis,* or key, began as a textbook on algebra said to be written to instruct the Earl of Arundel's sons. It contained in condensed form the essentials of arithmetic and algebra known at that time. Compared to other contemporary works, the *Clavis* utilised extensive symbolism and treated numbers in a sophisticated fashion. For example, Oughtred gave an explanation of decimal fractions and how to operate with them. He also accepted the idea of irrational numbers, and that they could be approximated with indefinite decimal expansions and continued fractions, and in several editions he included an exposition of logarithms. Thus in Oughtred's work we can clearly see the beginnings of a new concept of number and the continuation of the expansion of algebraic techniques into England.

The *Clavis* appeared in five Latin editions, the first in 1631, and the last in 1698. There were also two independent English editions, the first (translated by Robert Wood) in 1647, and the second in 1694. Few changes were made to the text by

Oughtred between these editions. I shall cite the 1647 English translation; having compared this text to the 1648 Latin edition I have found no appreciable differences.

Oughtred made it clear at the beginning of the *Clavis* that he was breaking new ground; he said his work was "not written in the usuall synthetical manner, nor with verbous expressions, but in the inventive way of *Analitice*, and with Symboles or notes of things instead of words."[79] This symbolic framework was meant to ease the understanding of ancient authors such as Euclid, Archimedes, Apollonius and Diophantus. Students could, using his method, go beyond learning the propositions of the ancients; they could now hope to understand how the Greek material was created. Oughtred admitted that he himself had difficulties following some of the more complicated arguments until he "uncasing the Propositions and Demonstrations out of their covert of words, designed them in notes and species appearing to the very eye."[80] Oughtred's claim was that he was providing a tool that would increase the understanding of existing mathematics, and that this tool might perhaps facilitate future discoveries.

He began his book by introducing Hindu-Arabic numerals, place notation, and decimal fractions. Although a great admirer of Napier, who had mastered the theory of logarithms soon after their publication in 1614, he did not follow Napier in using the decimal point or comma for designating decimals. Instead, he used a dot to express ratios. He considered symbols to be more applicable than numbers to the "Analytical art," or as he put it, the method "in which by taking the things sought as knowne, we finde out that we seeke."[81] In the chapters that covered addition, subtraction, multiplication and division, Oughtred explained these operations and gave examples using integers, decimals, and symbols.

Draw into	$3A$ $2A$	AE A	AE AE	$A+E$ $A+E$
Product	$6Aq$	AqE	$AqEq$	$Aq+AE$ $+AE+Eq$
				$Aq+2AE+Eq$

Figure 8. Oughtred's Operations

The small q here designates a square, so A times $A = Aq$. Although many of the examples given are algebraic, portions of the explanations have a geometric

[79] William Oughtred, *Clavis Mathematicae*, trans. Robert Wood (London: 1647), ff. B4i-B4ii.
[80] Oughtred, *Clavis*, f. B4iii. These notes and species refer to Oughtred's own brand of symbolism, which was similar to, but more complicated than, Viète's.
[81] Oughtred, *Clavis*, p.4

character. For instance, Oughtred referred to the number found in division as a "Parabola", or "application", because he believed "it ariseth out of the application of a plaine to a longitude given, to finde out a Latitude agreeing thereto." When discussing his equations the symbols are often meant to represent magnitude and their product a "rectangle contained under them."[82] Thus it appears that he was influenced by the Greek idea that the product of two line segments would be a

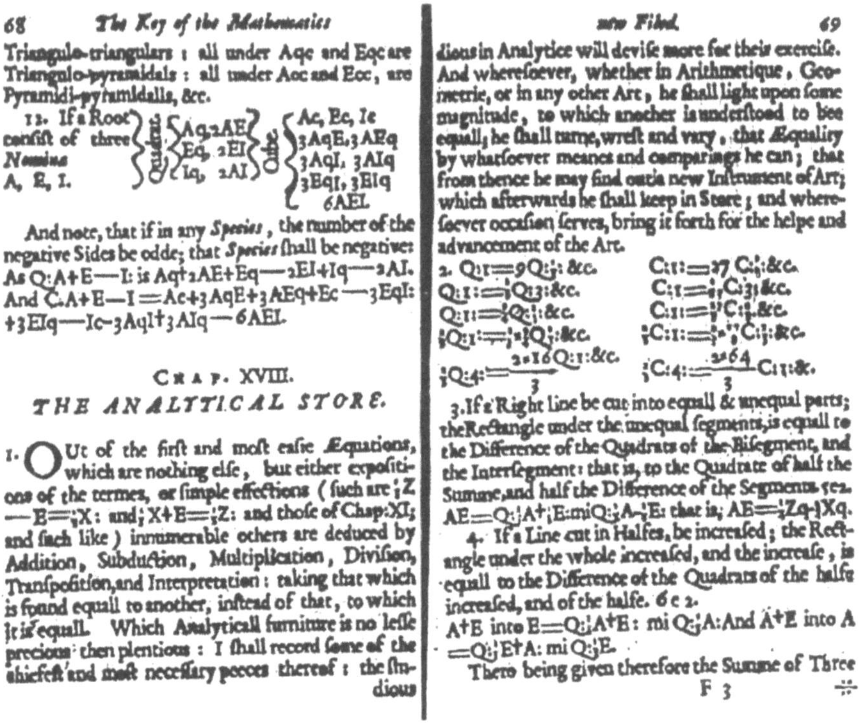

68 The Key of the Mathematicks

Triangulo-triangulars: all under Aqc and Eqc are Triangulo-pyramidals: all under Aoc and Eoc, are Pyramidi-pyramidalls, &c.

12. If a Root consiſt of three Nomina A, E, I. — Quadrat: Aq, 2AE; Eq, 2EI; Iq, 2AI. Cubic: Ac, Ec, Ic; 3AqE, 3AEq; 3AqI, 3AIq; 3EqI, 3EIq; 6AEI.

And note, that if in any *Species*, the number of the negative Sides be odde; that *Species* ſhall be negative: As Q:A+E—I: is Aq+2AE+Eq—2EI+Iq—2AI. And C:A+E—I = Ac+3AqE+3AEq+Ec—3EqI: +3EIq—Ic—3AqI+3AIq—6AEI.

CHAP. XVIII.

THE ANALYTICAL STORE.

1. OUt of the firſt and moſt eaſie *Æquations*, which are nothing elſe, but either expoſitions of the termes, or ſimple effections (ſuch are ½Z—E=½X: and ½X+E=½Z: and thoſe of Chap: XI; and ſuch like) innumerable others are deduced by Addition, Subduction, Multiplication, Diviſion, Tranſpoſition, and Interpretation: taking that which is found equall to another, inſtead of that, to which it is equall. Which Analyticall furniture is no leſſe precious then plentious: I ſhall record ſome of the chiefeſt and moſt neceſſary peeces thereof: the ſtudious

new Filed. 69

dious in Analytice will deviſe more for their exerciſe. And whereſoever, whether in Arithmetique, Geometrie, or in any other Art, he ſhall light upon ſome magnitude, to which another is underſtood to bee equall, he ſhall turne, wreſt and vary, that *Æquality* by whatſoever meanes and comparings he can; that from thence he may find out a new Inſtrument of Art; which afterwards he ſhall keep in Store; and whereſoever occaſion ſerves, bring it forth for the helpe and advancement of the Art.

2. Q:1:=9Q:⅓: &c. C:1:=27C:⅓:&c.
Q:1:=¼Q:2:&c. C:1:=⅛C:2:&c.
Q:1:=[illegible]Q:[illegible]:&c. C:1:=[illegible]C:[illegible]:&c.
⅓Q:1:=[illegible]Q:[illegible]:&c. ⅓C:1:=[illegible]C:[illegible]:&c.
⅓Q:4:=(2·16/3)Q:1:&c. ⅓C:4:=(2·64/3)C:1:&c.

3. If a Right line be cut into equall & unequall parts; the Rectangle under the unequall ſegments, is equall to the Difference of the Quadrats of the Biſegment, and the Interſegment: that is, to the Quadrate of half the Summe, and half the Difference of the Segments. ſez. AE=Q:½A+½E: mi Q:½A—½E: that is, AE=¼Zq—¼Xq.

4. If a Line cut in Halfes, be increaſed; the Rectangle under the whole increaſed, and the increaſe, is equall to the Difference of the Quadrats of the halfe increaſed, and of the halfe. 6 e 2.
A+E into E=Q:½A+E: mi Q:½A: And A+E into A =Q:½E+A: mi Q:½E.

There being given therefore the Summe of Three

F 3

Figure 9: Oughtred's style of equations

rectangle. But unlike such later mathematicians as John Wallis, Oughtred accepted division of both integers and "mixed" numbers or fractions by irrationals or what he called "infinite numbers." He also accepted that a unit could be divided as much as one liked and that the result was a number.[83] There is no evidence that he subscribed to the ancient distinction between discrete numbers and continuous magnitude, as Robert Recorde had. Moreover, he seems to have ignored the Aristotelian precept that one was not a number but an indivisible unit, and that the smallest number was two. Oughtred's thinking, although predominantly algebraic, was still tied to geometry. Nonetheless, in terms of his willingness to broaden

[82] Oughtred, *Clavis*, p. 33. In chapter twelve, Oughtred explains how to apply his symbols and their operations to familiar geometric problems.
[83] Oughtred, *Clavis*, p.28.

number concepts, he was genuinely advanced, willing to go further than later figures such as Wallis.

Having completed his explanation of basic arithmetic, Oughtred began a discussion of algebra. He used ordinary capital letters, *A, B, C, . . .* to designate given numbers.
The powers of aggregates were designated by prefixing the capital letter *Q* (for square), *C* (for cube), and by combinations of these capital letters for higher powers, such as *QQ* (for the fourth power).

Oughtred expressed aggregation by placing a colon (:) at both ends. Thus he wrote $Q : A - E :$ for $(A - E)^2$. He also represented in symbols certain composite expressions, such as $A + E = Z$, $A - E = X$, where *A* is greater than *E*. As we have seen, although Oughtred employed a great deal of symbolism, many of his explanations had a geometrical character. For instance, he described the powers made from roots or sides multiplied into themselves as being geometrical figures; he calls a "side drawne into it selfe" as a "Quadrat" and a "Quadrat" drawn into its side a "Cube". Indeed, unlike Recorde, Oughtred was careful to furnish a geometric background for algebraic operations.[84]

He also included a binomial table, and he also gave rules to expand the table "at pleasure" for positive integers. Oughtred then showed how one could use his binomial table to compute whatever power of a number that was desired. He also used his table to compute the roots of numbers. Some numbers, of course, did not yield exact roots, so he continued his computations past the decimal point (or in his case the dividing line) until he had the desired degree of accuracy.[85] Oughtred explained how to do arithmetic with, as he put it, "Commensurable Surde Sides," or numbers which when reduced to least terms "become true figurate numbers of their *Kinde*."[86] For example, $\sqrt{12}$ and $\sqrt{147}$ are commensurable surde sides because, when reduced to the least terms by their greatest common measure, $\sqrt{3}$, they become 2 and 7. But he was somewhat worried about homogeneity. He said that "Heterogeneal Sides are neither multiplied nor divided, unlesse they be first reduced to the same *Kinde*."[87] Oughtred seemed to have agreed to some extent with Viète about the importance of homogeneity constraints. The first edition of the *Clavis* was published in 1631, six years before Descartes' *Geometrie* suggested exponential notation and rejected homogeneity constraints, but there is no evidence that Oughtred was ever familiar with Descartes' work. Indeed, Oughtred's correspondences suggest that he had difficulty obtaining his contemporaries' mathematical texts from the Continent. For example, he heard of Cavalieri's methods in a letter from Paris, but he was never able to purchase the original book.[88] Oughtred seems to have been influenced mainly by ancient Greek mathematics and

[84] Carl Boyer, *A History of Mathematics* (Princeton: Princeton University Press, 1985), p. 337.
[85] Oughtred, *Clavis*, p.48.
[86] Oughtred, *Clavis*, pp. 49-50.
[87] Oughtred, *Clavis*, p.51.
[88] Stephen Rigaud, *Correspondence of Scientific Men of the Seventeenth Century* (Oxford: Oxford University Press, 1841), p.66.

Viète. In the introduction to the *Clavis*, he only specifically mentions the works of Euclid, Archimedes, Apollonius, Diophantus and Ptolemy.

After dealing with numbers and the rules of arithmetic Oughtred began to give rules for solving equations, including a nearly general solution of the quadratic equation and rules for simplifying equations such as transposition, division by the unknown, and division by a constant. Moreover, he believed that if one intended "to be an Analyst ...[one must] not be ignorant of these Geometricall Theorems and Problemes."[89] To support this viewpoint, Oughtred included all of the theorems in Euclid's *Elements* Book I in a very condensed form and without proof. This material seems to have been intended mainly as background information. He also showed how his approach with analytical equations could be applied to geometry by working through several of the problems in Euclid's *Elements* Book II using both algebraic and geometrical methods.

Oughtred gave an illustration of logarithms, a fairly new resource at this time, providing a detailed explanation and specific examples. Oughtred was familiar with logarithms because Henry Briggs had consulted him while Briggs "was perfecting Napier's work".[90] His exposition of logarithms, however, was not included until the 1647 edition of the *Clavis*. The treatment of the rules for operating with logarithms, which is brief, is similar to modern practices. For instance, he explains that "The Summe of two Logarithmes is the Logarithme of the Product of their Values; and the difference is the Logarithme of the Quotient."[91]

Oughtred ended the 1647 edition with two chapters of mercantile-type problems. He explained how to compute compound interest. Several of the examples deal with such practical issues as profits on common stocks and theorems to calculate the income of pensions. He included a table of financial formulae.[92] Oughtred's general interest in practical mathematics can also be seen in the effort he put into designing mathematical instruments and surveying instruments. He was also very interested in the construction of sun-dials. Oughtred was the first inventor of slide rules around 1622, although his description of his instruments was not in print before 1632.[93] But there was always a certain tension in Oughtred's work about the relative merits of theory and practice. For instance, although he took the trouble to design instruments, he always spoke in depreciating terms of their importance and of their educational value.[94]

The *Clavis* contains no startling mathematical discoveries, being more of a summary of the algebra known in England in this period. It was, however, considered by Oughtred's contemporaries to be a difficult book. Several of his

[89] Oughtred, *Clavis*, p.72.
[90] Cajori, *William Oughtred*, p.6.
[91] Oughtred, *Clavis*, p.135.
[92] Oughtred, *Clavis*, p.172.
[93] Cajori, *William Oughtred*, p.47.
[94] For more details on Oughtred's views on this subject, see Katherine Hill "'Juglers or Schollers?': Negotiating the Role of a Mathematical Practitioner" *British Journal for the History of Science* **31** (1998): 253-74.

friends wrote letters to Oughtred begging him to expand his text. William Robinson wrote,

> I will once again earnestly entreat you, that you be rather diffuse in the setting forth of your English mathematical *Clavis*, than concise . . . these arts cannot be made too easy, they are so abstruse of themselves, and men either so lazy or dull, that their fastidious wits take a loathing at the very entrance of these studies, unless it be sweetened on with plainness and facility . . . .[95]

Oughtred, on the other hand, believed that students with enough mathematical talent, who studied his book carefully, needed no other aid.[96] Some of his readers, such as Seth Ward, were said to have so much difficulty following the *Clavis* that they went to live with Oughtred and received free instruction in mathematics.[97] He also answered many letters which requested further explanations or help in solving particular problems. Indeed, his reputation as a mathematics teacher was so great that men he did not know wrote to him for help.[98] For instance, his friend, Dr. Twyden, wrote to him that "I think it little less than robbery to keep back from you this knowledge of that estimation [which] is put upon your person and book in foreign parts . . ."[99] Thus, in his day, Oughtred had a widespread reputation as a mathematician and a teacher, although the *Clavis* was later overshadowed by Wallis' work.

Oughtred's *Clavis* was very much a transitional book of algebra. In it, algebra did not rise above the limitations imposed by geometric thinking. But Oughtred did ignore many classical strictures about numbers in order to implement his analytic techniques. As Klein suggests, the increased use of algebraic symbolism is in Oughtred's case found together with a broadened number concept. Oughtred treated irrationals, decimal expansions, fractions and logarithms as numbers, and he even accepted the existence of negative numbers, although not as the roots of equations. In effect, he treated numbers as continuous, rather than discrete, which was a movement away from Greek concerns. The relationship of algebra to geometry, however, remained complex. Oughtred treated algebra similarly to Viète, as a special procedure for discovery. His concern with homogeneity shows that Oughtred's thoughts, like Viète's, were always close to geometry. Moreover, his rejection of negative, or imaginary roots, might have been due to the lack, in this period, of a geometrical interpretation. In spite of its ambivalent character, it had a great impact on the development of mathematics: John Wallis, Isaac Barrow, and Isaac Newton were all influenced by it.

CONCLUSION

As Klein suggests, all of the algebras examined show a movement towards conceiving numbers as symbols insofar as variables were being used instead of exact

---

[95] Stephen Rigaud, *Correspondence of Scientific Men of the Seventeenth*, p.27

[96] Rigaud, *Correspondence of Scientific Men of the Seventeenth*, p.10.

[97] Cajori, *William Oughtred*, p.60.

[98] Rigaud, *Correspondence of Scientific Men of the Seventeenth*, p 59.

[99] Rigaud, *Correspondence of Scientific Men of the Seventeenth*, p. 67.

numerical examples. As we have seen in our examination of the arithmetics, practical problems alone, without the use of symbolism and approximations techniques, do not seem to broaden number use beyond positive fractions. The early modern English arithmetic and algebraic tradition support Klein's claim the conceptual transformation in number concepts was heavily influenced by the development of algebraic techniques. Oughtred, in particular, both stressed in importance of the symbolic framework and went the furthest in stretching number concepts. What would be accepted as an appropriate value for the variable seems to have been heavily influenced by the types of problems that the mathematicians were trying to solve. In many of Recorde's problems the solution was usually a positive integer. It would not have made sense to attempt to build a wall with a negative number of bricks. However, in some of the commercially oriented problems rational solutions could be depicted in a manner that made common sense insofar as the answer could be thought of as so many pounds, shillings, and pence. In cases where a measurement was needed, some answers were given in terms of approximations of positive irrationals. Oughtred's problems were less likely to be modelled on practical activities such as building, measuring, or commerce, although at the end of the *Clavis* he did include commercial examples dealing with interest. His intent was rather to give mathematicians a tool that would help them to better understand traditional Greek mathematics, and that could also be used to expand this knowledge. But not all the solutions to his equations were positive integers. In the end Oughtred was willing to treat irrational, decimal expansions, fractions and logarithms all as being numbers. He accepted all solutions that could be given a geometrical interpretation as numbers. Harriot's mathematics was usually motivated by physical considerations. Although he realised that negative and complex roots were possible solutions, they were not useful for solving problems in free fall and optics. Thus, although he realised the mathematical possibility of these solutions, he did not view them as having utility in actual practice. In his work on navigation and ballistics, on the other hand, he used broadened number concepts and numerical approximations in order to solve problems.

Although these scholars were familiar with Aristotle's dictates on the nature of numbers, and they often made reference to his concerns, when they needed the answers to actual problems they used the numbers they needed. The nature of these "actual problems" changed over time: for Recorde they were usually taken from practical activities such as measuring or war; for Oughtred they might be the expression of a quadratic equation that he believed captured the essence of a Greek geometric problem; Harriot was often interested in mathematics that would further his research in what we would now call physics. None of these mathematicians shows signs of being overwhelmed with ontological worries about the true nature of numbers. In some cases they were willing to treat numbers as though they were continuous, even though there was some hesitation about labelling fractions and irrationals as proper numbers. The concern, shown by Recorde and Oughtred, to invent labels for numbers that could not be expressed as positive integers might indicate worries as to their actual status. On the other hand, the results they obtained were often too useful to ignore. There was a certain amount of tension between the realisation that there was no rigorous basis for treating numbers as continuous and the desire to obtain useful results. It was only when the answers could not be placed inside any kind of meaningful framework, as was the case with complex numbers, that numerical results were ignored.

CHAPTER 5

THE DEVELOPMENT OF THE LOGARITHMS

Napier and Briggs

The invention of logarithms in the late sixteenth century by John Napier and the further development of his work by Henry Briggs had a profound impact on conceptions of abstract magnitude and number and the overall evolution of mathematics in early modern England and Scotland. Napier invented new mathematical techniques; he utilised points moving through space and concepts that we would now call velocity and acceleration in constructing his model. Indeed, the concepts of time, motion and instantaneous speed allowed Napier to model a correspondence between an arithmetic and a geometric progression. Henry Briggs, on the other hand, continued Napier's work on the logarithms and converted them to the more familiar "base ten system" using solely numerical methods. Briggs was perhaps first to use the calculus of finite differences to construct a table of logarithms of base ten. In the course of his work on logarithms, Briggs was also the first to develop the binomial theorem, $(a + b)^n$, for a non-integer values of n. Although they were working on the same subject, the difference between Napier's and Briggs' approach is quite striking.

Napier founded his version of the logarithms on a kinematic conception of magnitude while Briggs adopted only numerical techniques. Even though they both utilised decimal expansions and treated numbers as though they were continuous, or a type of abstract magnitude, with the only difficulty being the degree of precision desired, their underlying methods were very different. Napier took the view that these logarithmic numbers could be given a proper foundation by mapping them to continuous geometric magnitude using kinematic methods. Briggs, however, seemed to feel no need to justify his numerical methods. He ignored Napier's foundation and worked directly with numerical techniques.

These two approaches, of kinematic and numerical techniques, were to reappear in English mathematics. Isaac Barrow was later to use kinematic methods similar to Napier's in his *Geometrical Lectures*, and his views about the relation between number and magnitude seem parallel to those of Napier as well. John Wallis, however, concentrated on developing numerical methods comparable to Briggs'. Wallis, like Briggs, attempted to develop the binomial theorem for non-integer exponents.

Logarithms were always depicted in terms of decimal expansions. They were another new kind of number that could not be confined to the positive integers. As well, Napier's approach directly associated number and geometric magnitude. Although the developers of the logarithms did not utilise algebraic symbolism, they indirectly treated numbers symbolically, and as though they were interchangeable with geometric magnitude. The primary motivation for the creation of logarithms was the desire to more easily solve problems associated with navigation and the trades.[1] By the time of Briggs' death, they were also being used in commerce. The logarithms not only played a significant role in the broadening of number concepts and the changing views of the relationship between number and magnitude, but they also show how a need to solve practical problems can lead to new mathematical ideas.

In Napier's scheme, logarithms were numbers that were considered to be labels for the measure of a magnitude. He seemed to view a type of number he labelled "concrete" in his work on algebra as being associated with continuous magnitude.[2] Napier did not, of course, have a well-developed notion of real numbers. He used approximations of irrationals to his desired degree of accuracy, but he did not seem overly concerned with their ontological status. Napier acted as though every point on a line could be assigned a number, even if that number could only be approximated and not computed exactly.

The development of the logarithms is connected to practical concerns; the nature of the problems influenced treatments of number and ideas about its relationship to magnitude. The number of calculations necessary to complete the tables needed by astronomers in this period was truly daunting, and Napier's goal was to simplify trigonometric calculations. Trigonometry was an independent field of study used in astronomy, surveying, map making, and navigation, and much time was devoted to calculating extensive tables of sines and tangents. Napier's logarithms were logarithms of sines because they were intended to be useful in trigonometric calculations. Safe and fast navigation was of the utmost importance during this period in Britain, and trigonometry was vital to problems in this area.[3] Briggs was also interested in these issues.

Napier's goal was similarly practical, he wished to replace the operations of multiplication, division, and extraction of roots of sines and tangents with the operations of addition, subtraction, and division by two. Briggs's tables made computations even easier, because in his table the logarithm of one was zero. As well, they had wider applications than Napier's tables because rather than being defined in terms of sines, Briggs' logarithms were logarithms of numbers using the more convenient base ten.

Napier could have computed the logarithms and therefore simplified arithmetic calculations without using kinematic concepts. Joost Bürgi, a Swiss maker of watches

---

[1] David Waters, *The Art of Navigation in England in Elizabethan and Early Stuart Times* (London: Hollis and Cauter, 1958). See chapter six, "Navigational Practice: 1623-1632" for details concerning the importance of logarithms to "scientific" navigation.

[2] W. F. Hawkins supplies a loose translation and commentary of this work in his PhD thesis "The Mathematical Work of John Napier," (University of Auckland, 1982). The 1839 volume is very rare as only 101 copies were printed for members of the Maitland Club. The manuscript *De Arte Logistica* was derived from was the only of of Napier's papers to survive. All of Napier's papers had been left to his son Robert, when the house of Napier was burnt, all the other papers perished.

[3] Herman H. Goldstine, *A History of Numerical Analysis from the 16th through the 19th Century* (New York: Springer-Verlag, 1977), p.13.

and astronomical instruments, independently calculated logarithms at about the same time as Napier after noting a relation between geometric and arithmetic progressions using solely numerical considerations. Bürgi, however, did not publish his work until 1620, and Napier does not seem to have known about this independent development of the logarithms.[4]

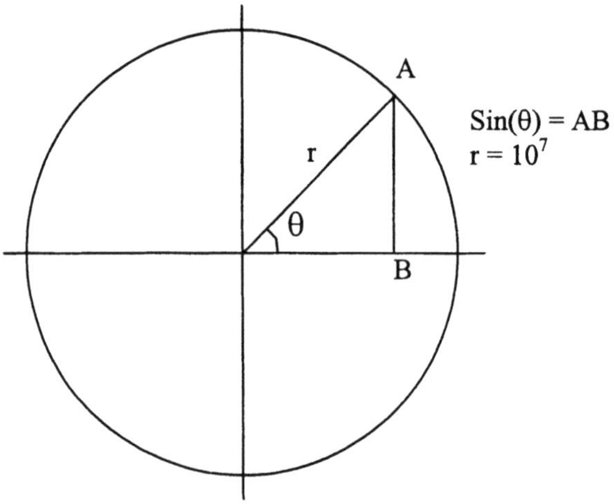

Figure 1.

Why did Napier take a kinematic approach? Lines traced by points moving through space were historically considered to be continuous. If Napier then associated the points on the lines with numbers he would, intuitively at least, have guaranteed the continuity of the numbers. Strictly speaking, however, the velocity of the moving points he utilised varied continuously and a rigorous expression of his model would have called for the use of calculus. Moreover, the model was intuitive in another way. Napier was dealing with the logarithm of sines, and in this period sines were thought of as the length of a semi-chord of a circle of a given radius subtending a specified angle at the centre. One of the moving points was designed to represent the whole sine decreasing to the specific sine of interest as the right angle decreased. In other words, at the beginning of the motion of the point associated with the sine line, the angle θ could be considered to be decreasing, and the line representing the whole sine would decrease; one could halt the decrease of the angle, and so the motion of the point, at the particular sine of interest. The use of the concepts of velocity and acceleration allowed Napier to locate the points of interest on the line, or in other words, they allowed him to calculate a specific location. This was important in that it allowed him to map at any

[4]Margaret Baron, "Napier, John," *Dictionary of Scientific Biography*, p. 610.

given time a point on the sine line to a point on the line representing the logarithms. Since Napier had no tools for handling what we would now consider to be a continuous function, he needed an intuitive basis.

In this chapter, I will first briefly discuss Napier's background and his mathematical activity prior to his invention of the logarithms. This will provide us with Napier's earliest opinions concerning numbers and algebra. Second, I will describe Napier's invention of the logarithms and their historical context. We will then be able to explore how logarithms influenced perceptions about numbers and helped to broaden the number concept. I will also consider the entirely different approach to the logarithms taken by Briggs. Moreover, an examination of a posthumous English translation of one of Briggs's texts will show how quickly logarithms were adapted to a variety of practical pursuits. The logarithms, therefore, helped to spread the broadening number concept into the practically oriented, non-Latin reading community.

XIII. PROP. *Hauing the Latitude and Longitude of a Starre; to finde his Declination and right Aſcenſion.*

For illuſtration of this Queſtion, let O be the Rams head, whoſe Longitude B N in the yeare 1630 is according to the calculation of *Tycho Brahe*, 2 gr. 32 min. of ♉, and his Latitude O N 9 gr. 57 min. Northwards, we would know his Declination O L, and his right Aſcenſion D L.

Here muſt be reſolued the ſphericall Triangle O E F, whoſe ſide O E is 80 gr. 3 min. (the Complement of the Latitude O N 9 gr. 57 min.) the ſide E F 23 gr. 30 min. (the diſtance of the Poles) and the angle O E F 57 gr. 28 min. (the Complement of the Longitude from the Spring Æquinoxe) and the ſide F O (the Complement of O L the Declination) is to be ſought after the manner of the 23 *Prop.* of the 10 *Chapter*, and the angle E F O (repreſented by D L the arch of the Æquinoctiall) by the 22 *Prop.* of the ſame *Chapter*, for the right Aſcenſion.

9,73061,29	the Co-ſine of 57 gr. 28 min. for O E F.
10,75590,28	the Tangent of 80 gr. 3 min, for E O.
10,48651,57	the Tangent of 71 gr. 56 min. for E P the firſt arch found.
9,49153,45	the Co-ſine of 71 gr. 56 min. the firſt arch found.
9,23751,53	the Co-ſine of 80 gr. 3 min. for E O.
9,82182,50	the Co-ſine of 48 gr. 26 min. for F P the ſecond arch found.
19,05935,03	the ſumme.
9,56781,58	the Co-ſine of 68 gr. 18. min. for F O, and therefore 21 gr. 42 min. for O L the Declination required.
9,97804,18	the Sine of 71 gr. 56 min. the firſt arch found.
10,19525,53	the Tangent of 57 gr. 28 min. for O E F.
20,17329,71	the ſumme.
9,87400,85	the Sine of 48 gr. 26 min. the ſecond arch found.
10,29928,86	the Tangent of 116 gr. 39 m. for E F O the angle in the Æquinoctiall from the winter Solſtice, and thereof being ſubtracted 90 gr. the remainder 26 gr. 39 min. is the right Aſcenſion required, which is 1 houre 46 min.

Now followes the names of ſome principall Stars with their Declination, diſtance from the Pole, and right Aſcenſion in Houres and Minutes, according to the calculation of *Tycho Brahe* for the yeare 1630, and may ſerue for many yeares after.

Figure 2. Briggs' use of Logarithms to solve a problem about navigation.

NAPIER'S BACKGROUND

John Napier (1550 - 1617), the eighth laird of Merchiston, was the son of Sir Archibald and Janet Bothwell. He was a descendent of a very well off family who owned lands in Brairnavaidis, the Isle of Inchmone, the lands of Edinbelle and Gartness in Stirlingshire, the Putrielands, and a house at Bairnisdale among other properties.[5] When Napier was thirteen he was sent to the University at St. Andrews to the College of St Salvator, but he remained there for only one or two terms. He may thereafter have studied abroad. By 1571, he had returned to Scotland.

He seems to have begun his first mathematical work sometime before 1573: a treatise on algebra that was not published in his lifetime. His contemporary fame, however, rested upon his first publication in 1593, *A Plaine Discovery of the Whole Revelation of St. John*, a damning indictment of Catholicism that pictured the Pope as the Antichrist. This work gave him a great reputation as a scholar and a theologian, and it was translated into Dutch, French and German, going through several editions in each language.[6] In this period Scotland was still troubled by the Armada scare, and Napier sent off a proposal for weapons to the English government in 1596. He developed a burning mirror to burn enemy ships at a distance, a piece of artillery for destroying everything within the arc of a circle and an "armoured chariot" that could fire in all directions. It is uncertain if any of these machines was ever constructed.[7] He also developed agricultural techniques for the improvement of crops and cattle. Napier's interest in a broad range of practical problems probably influenced his decision to work in an area of mathematics that had numerous real-world applications.

His interest in simplifying trigonometric calculations developed after learning of the problems Tycho Brahe was having with his astronomical calculation through a doctor, Thomas Craig, who had met Brahe in Denmark.[8] Kepler wrote a letter in 1624 to his friend Cugerus in which he stated, "But nothing, in my mind, surpasses the method of Napier, although a certain Scotchman, even in the year 1594, held out some promise of the wonderful Canon in a letter to Tycho."[9] This is generally taken to mean that Napier began working on logarithms by 1594.[10] Napier's first work on this subject, published in 1614, was his *Description of the Admirable Canon of Logarithms*. Edward Wright, an Oxford mathematician, was immediately hired by the Right Worshipful Company of the Merchants of London trading to the East Indies to translate Napier's Latin work into English for the benefit of the navigators and trainees who did not read Latin.[11] This

---

[5] Francis Shennan, *Flesh and Bones* (Edinburgh: Napier Polytechnic, 1989), p. 24.
[6] Shennan, pp. 62-69 and Baron, "Napier, John, *Dictionary of Scientific Biography,* p. 609. See also the catalogue of the works of John Napier attached to William Rae Macdonald's 1889 translation of *The Construction of the Wonderful Canon of Logarithms*, pp. 109-129, for full references to every edition.
[7] Shennan, pp.84-85.
[8] Shennan, p. 110.
[9] Quoted in Florian Cajori, "Algebra in Napier's Day and Alleged Prior Inventions of Logarithms," *Napier Tecentenary Memorial Volume*, ed. Cargill Gilston Knott (London: Longsmans, Green and Co., 1915), p. 104.
[10] E.W. Hobson, *John Napier and the Invention of Logarithms 1614* (Cambridge: Cambridge University Press, 1914), p. 13.
[11] William Hawkins, "The Mathematical Work of John Napier (1550-1617*)"* diss., University of Auckland,

translation was printed in 1616, with Napier's complete approval, and its dedicatory poem praises the use of logarithms in "true Measuring of Lands, Plots, Buildings and fortifications, Astronomie and Dialling, Geography and Navigation."[12]

Napier's second book on logarithms, *Constructio Canonis Logarithmorum*, was published in 1619. The *Constructio* was actually written several years before the *Descriptio.*[13] Napier died before the *Constructio* was ready to be published and his son Robert and Henry Briggs completed the work. This text presents Napier's original conception of logarithms and the way in which the tables had been calculated. But before I turn to Napier's development of the logarithms, I will examine one of his other mathematical works in order to provide a clearer picture of his conception of numbers.

ALGEBRA

Many mathematicians were employed during Napier's lifetime to compute the tables of astronomy and navigation that were needed after Tycho Brahe showed that the Alfonsine tables and those of Regiomantanous were unreliable. Napier wanted to reduce the burden of lengthy calculation, but he first had to master all the known techniques of the calculators. We can trace Napier's early mathematical work though a manuscript that, although not ready for publication, was copied by his son Robert for Henry Briggs. The original manuscript was lost in a fire, but a copy of the work was published in 1839 under the title *De Arte Logistica.*[14]

This work supplies us with Napier's views on numbers and algebra prior to his invention of logarithms. It seems to have been assembled soon after Napier's return from the continent in 1572.[15] It is unlikely that Napier learned any of his mathematics in Scotland; no arithmetic, geometry or algebra was taught in Scottish grammar schools. The universities taught only arithmetic of the scholastic type and the rudimentary elements of geometry from Boethius. Napier did attend the University of St. Andrew's at the age of thirteen, but he stayed for only one or two terms. After he left St. Andrew's he might have studied in France, the Low Countries, or Italy.[16] He may also have learned some mathematics during his travels.

One striking feature of Napier's work on algebra is the lack of symbolism-addition was indicated by "et" or "adde" and 3 times 5 was written as "3 per 5." Although he did not use cossic notation, it is possible that he was familiar with Recorde's *The Whetstone of Witte*, as he did utilise Recorde's symbol for equality.[17] He made up, however, his

---

1982, p. 40.

[12] John Nepair, *A Description of the Admirable Table of Logaritmes* (London: 1616), published in facsimile by Da Capo Press (New York: 1969). Napier's name was spelled in several different ways in this period and I have therefore decided to use in the text the version most familiar to modern readers.

[13] Lord Moulton, "The Invention of Logarithms its Genesis and Growth," *Napier Tecentenary Memorial Volume*, ed. Cargill Gilston Knott (London: Longsmans, Green and Co., 1915), pp. 3-6.

[14] J. E. A. Steggall, "A Short Account of the Treatise *De Arte Logistica*," *Napier Tecentenary Memorial Volume*, ed. Cargill Gilston Knott (London: Longsmans, Green and Co., 1915), p. 145.

[15] Baron, *DSB*, p.609.

[16] Francis Shennan, *Flesh and Bones* (Edinburgh: Napier Polytechnic, 1989), p.27.

[17] Interestingly, Napier was familiar with cossic notation, and he even supplied a table explaining the different

own notation for roots. Napier was also familiar with the idea of index notation, for example, he wrote 7(III) for 7^3. Overall, the book seems to be a step backwards from Recorde's work in terms of notation.[18]

The first book dealt with arithmetic, describing basic calculations with addition and subtraction, calculations of powers and roots, and proportions. Napier seems to have acknowledged both positive and negative whole numbers as well as fractions. Negative numbers were said to be less than zero and carry the minus sign before them; he explained their meaning in terms of debt.[19] The origin of fractions was said to be in division; he defined fractions as numbers that have a denominator, differing from unity, placed beneath the numerator. Napier was careful to point out that it is impossible to find any common factor for a surd, or irrational number. He stated, "here I must warn you strongly against dividing incommensurable numbers which can never terminate. . . .For the square root of the number 10 ... no common measure can ever be found."[20] He was willing, however, to give approximate values for roots.

Napier made an interesting distinction between pure and applied mathematics. He considered arithmetic to be "the Logistic of discrete quantities by the use of individual numbers."[21] Fractions were included under discrete numbers because they "measure out its [number] own unique and individual ranking."[22] This is an unusual point of view, as fractions were not normally considered to be discrete. Pure mathematics, for him, appears to have only utilised discrete numbers. Approximations, on the other hand, "are suited to Applied Mathematics rather than to pure mathematics because imperfect roots cannot be evaluated exactly and give approximate results . . . ."[23] He believed these methods were suited more for the construction of sine tables or for the use of builders. The classical division of mathematics into arithmetic and geometry and the Aristotelian distinction between pure and applied mathematics seem to have been accepted by Napier. Pure mathematics dealt with the intrinsic properties of number and magnitude, and applied mathematics included such things as music and astronomy.

The second book moved beyond arithmetic into algebra. In this book Napier continued his distinction between types of numbers. He stated that "geometry is the logistics of material quantities by the use of concrete numbers. . . .A number is called a concrete number insofar as it relates to a material and continuous entity."[24] All

---

symbols. He also discussed other popular notations for roots used at this time. His own notation remained, of course, unpublished.

[18] W. F. Hawkins supplies a loose translation and commentary of this work in his PhD thesis "The Mathematical Work of John Napier," (University of Auckland, 1982). The 1839 volume is very rare as only 101 copies were printed for members of the Maitland Club.

[19] Hawkins, p.15, and J.E.A. Steggall, "A Short Account of the Treatise *De Arte Logistica,*" *Napier Tecentenary Memorial Volume*, ed. C.G. Knott (London: Longsmans, Green & Co., 1915), p.148.

[20] Hawkins, p. 20.

[21] Hawkins, p.25.

[22] Hawkins, p.76.

[23] Hawkins, pp. 61-63.

[24] Hawkins, p.80. Steggall gives a different definition of a concrete number: first, it must refer to a real quantity and second, it must be an incommensurable root of a number. The *Oxford English Dictionary* (Oxford: Clarendon Press, 1961) says that "concrete" was applied by the early logicians and grammarians to

irrational numbers were labelled concrete because no fractional numbers are commensurable with them. It would seem that Napier might have also considered logarithms to be concrete numbers as they related to a material and continuous entity. Unfortunately, Napier's son Robert could not find much of the material relating geometry to algebra and numbers, so all that remains is a few tantalising hints.

Napier defined algebra as the science that leads to the solution of the questions "how many?" and "of what kind?" He again took up surds, or irrational numbers, as being the roots of rational numbers that have no roots among the "absolute" numbers. Moreover, he considered them to be in some sense physical quantities that linked them to geometry.[25] He claimed that what could be done with rational fractions in arithmetic could be accomplished for irrational fractions by algebra. However, he gave few specific examples in the sections actually dealing with algebra.

Napier, like Recorde, discussed cossic numbers. He stated, "Those numbers, whether rational or irrational which bear the cossicke symbol, are called cossicke numbers. For example, *6R, 5a, 7bC*. . . are called cossicke numbers."[26] He used all the letters of the alphabet to represent his unknowns, and the usual symbol "Q" for numbers multiplied by themselves. Next he explored operations such as addition, subtraction, multiplication and division with these cossic numbers. The solutions to examples were given rhetorically. The final chapters discussed simple algebraic equations, which were described in terms of proportionals. The solutions given are similar to those supplied by Recorde, although Napier seems to have anticipated Harriot's idea of placing everything on one side of the equation and setting it equal to zero.

Napier's work on arithmetic and algebra prior to his invention of the logarithms shows that he was willing to accept fractions, irrationals, and negative quantities as numbers. Although he maintained the traditional distinction between pure and applied mathematics that led him to place various numbers into different categories that were partially dependent upon their applications, he freely operated on the entire range of real numbers. Moreover, it is unclear that he was particularly bothered by the idea of numbers being defined as collections of units. His description of his reasons for his invention of logarithms makes it clear that his greatest concern was ease of calculation and generating enough accuracy for applications.

NAPIER'S LOGARITHMS

Archimedes, in his *Sand Reckoner*, was aware of the idea of a geometric progression such as Napier used to develop the logarithms. He also had a theorem that indicates that he understood the relationship

---

a quality viewed as *concreted* or adherent to a substance, and so the word expressed a quality. In mathematics, seven (men or days) is a *concrete number* as opposed to the number 7 in the abstract. In 1594 Blundeville stated, "Understand, that of numbers some are said to be abstract, and some concrete." p. 776.

[25] Hawkins, p.86.

[26] Hawkins, p. 142.

$$a^m \cdot a^n = a^{m+n}.^{27} \tag{1}$$

Moreover, Archimedes discussed arithmetical progressions in terms of magnitudes in his work *On Conoids and Spheroids*.[28] Napier might have been influenced by Archimedes's use of points moving through space to construct curves in his work *On Spirals*, and since Archimedes' works had been translated into Latin and published by the 1540s, Napier could have been inspired by his kinematic approach.[29] He at least mentions both Archimedes and Euclid in his work, but he does not cite any of the relevant contemporary European authors.

Long before Napier's invention of the logarithms some European texts had accepted negative solutions to problems, given progressive approaches to approximating irrationals by a sequence of rational fractions, and begun work on exponents.[30] In 1484, Nicolas Chuquet wrote *Le Triparty en la Science des Numbers*, a work that used negative numbers as both coefficients and as exponents, allowed negative solutions to equations, and considered negatives, surds, and zero as numbers. Indeed, Chuquet had a very general concept of number.[31] This work also expresses very clearly the relations,

$$a^m a^n = a^{m+n} \tag{2}$$

$$(a^m)^n = a^{mn}.^{32} \tag{3}$$

Michael Stifel's *Arithmetica integra*, published in 1544, also includes a discussion of exponents and irrationals. Stifel's work provides insight as to the status of irrationals in this period. In the beginning of his second book, he wrote about his study of Euclidean irrationals, reminding his readers that there was a disagreement as to whether they were "true" numbers or not. Moreover, he stated that irrationals can not be expressed with precision and that they do not bear a definite ratio to "true" numbers.[33] Thus although many works utilised irrationals, many mathematicians were still unclear about the numerical status of irrationals. Stifel, in this work, used the series

---

[27] T.L. Heath, *The Works of Archimedes* (New York: Dover, 1897), pp. 220-230.

[28] Heath, *Archimedes*, p.105.

[29] J. G. Schonbeck, "Euclidean and Archimedian traditions in the Middle Ages and the Renaissance," *Companion Encyclopedia of the History and Philosophy of the Mathematical Sciences* (London: Routledge, 1994), pp.182-83.

[30] Jacques Sesiano, "On an Algorithm for the approximation of Surds from a Provencal Treatise," in *Mathematics from Manuscript to Print: 1300-1600*, ed. Cynthia Hay (Oxford: Clarendon Press, 1988) describes a copy of an original text by an unknown author from around 1430 that gives an algorithm for the approximation of "surds" to any degree of accuracy that also accepts negative numbers as solution to problems.

[31] Graham Fleg, "Nicholas Chuquet--an Introduction," *Mathematics from Manuscript to Print: 1300-1600*, ed. Cynthia Hay (Oxford: Clarendon Press, 1988), p.60.

[32] David Smith, "The Law of Exponents in the Works of the Sixteenth Century," *Napier Tercentenary Memorial Volume*, ed. C.G. Knott (London: Longsmans, Green & Co., 1915), p.83.

[33] Sesiano, p.39.

0	1	2	3	4	5	6	7	8
1	2	4	8	16	32	64	128	256

calling the upper numbers *exponents*. Moreover, he stated four rules: addition in arithmetical progression corresponds to multiplication in geometrical progression; subtraction corresponds to division; multiplication to the finding of powers; and division to the extraction of roots.[34] Stifel also used negative exponents. Although Stifel knew many of the concepts we normally associate with logarithms, Napier took an entirely different path.

Napier approached logarithms through considerations that are quite geometrical, or kinematic in that he utilised the concept of the velocity of flowing points. We will begin our exploration of his methods through a study of his *A Description of the Admirable Table of Logarithms*, since it was his first work published on logarithms and it had the most influence on their future development.

Napier described his motivation in the preface to the *Descriptio*. He stated

> There is nothing (right well beloved Students in the Mathematicks) that is so troublesome to Mathematicall practice, nor that doeth more hinder Calculators, then the Multiplications, Divisions, square and cubical Extractions of great numbers, which besides the tedious expense of time, are for the most part subject to many slippery errors. I began therefore to consider in my minde, by what certaine and ready Art I might remove those hinderences...[35]

Napier's work was specifically directed towards simplifying astronomical calculations. In general, logarithmic tables were produced in response to the needs of the practical mathematicians and astronomers of the time to reduce the time consuming and error prone arithmetical procedures. The *Descriptio*'s text consists of two books. The first book contains definitions and the working rules of logarithms and the second book applies logarithms to various problems of trigonometry, both plane and spherical. By the end of the first six definitions of the first chapter, Napier had introduced logarithms.

The first definition stated that a motion of a point that covers equal spaces in equal times creates an infinite line, A_i that can be said to increase equally.[36]

---

[34] Smith, p. 86 and Morris Kline, *Mathematical Thought From Ancient to Modern Times*, Vol. 1 (Oxford: Oxford University Press, 1972), p.256.

[35] John Nepair, *A Description of the Admirable Table of Logarithms* (London: 1616), published in facsimile by Da Capo Press (New York: 1969).

[36] Nepair, *A Description*, p.1.

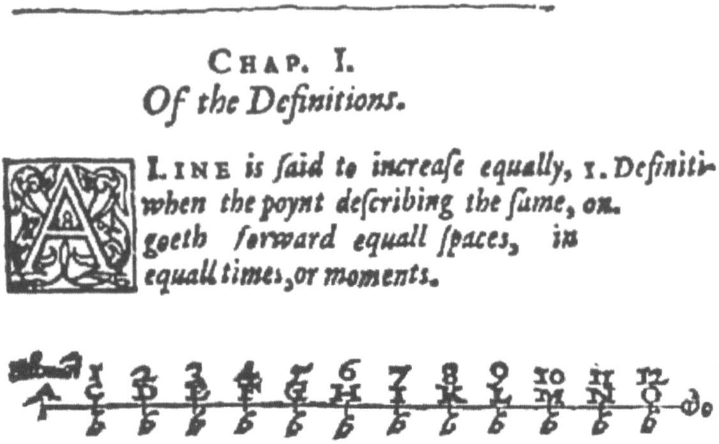

Chap. I.
Of the Definitions.

A Line is said to increase equally, when the poynt describing the same, goeth forward equall spaces, in equall times, or moments. 1. Definiti-on.

Figure 3.

He then stated a corollary that "Therefore this increasing, quantities equally differing, must needes be produced, in times equally differing."[37] These concepts are similar to those used by Oresme and the Merton School, and later by Galileo in his 1638 work *Dialogues Concerning Two New Sciences*. On the third day Galileo gave definitions dealing with uniform motion, he stated "By steady or uniform motion, I mean one in which the distances traversed by the moving particle during any equal intervals of time, are themselves equal."[38] Galileo later provided an axiom that serves the same purpose as Napier's corollary: "In the case of one and the same uniform motion, the distances traversed during a longer interval of time is greater than the distance traversed during a shorter interval of time."[39] Thus, the kinematic methods used by Napier were also utilised by some of the creators of the 'new physics' as well as by later mathematicians such as Barrow.

Napier's second definition dealt with nonuniform acceleration. He stated,

> A Line is said to decrease proportionally into a shorter, when the poynt describing the same in equall times, cutteth off parts continually of the same proportion to the lines from which they are cut off.[40]

In his later work he likened this process to a geometric series. He used a series of diagrams to make his meaning clear.

[37] Nepair, *A Descripition*, p.2.
[38] Galileo Galilei, *Dialogues Concerning Two New Sciences*, trans. Henry Crew and Alfonso de Salvio (New York: Dover, 1954), p. 154.
[39] Galileo, p.154.
[40] Nepier, *A Description*, p. 2.

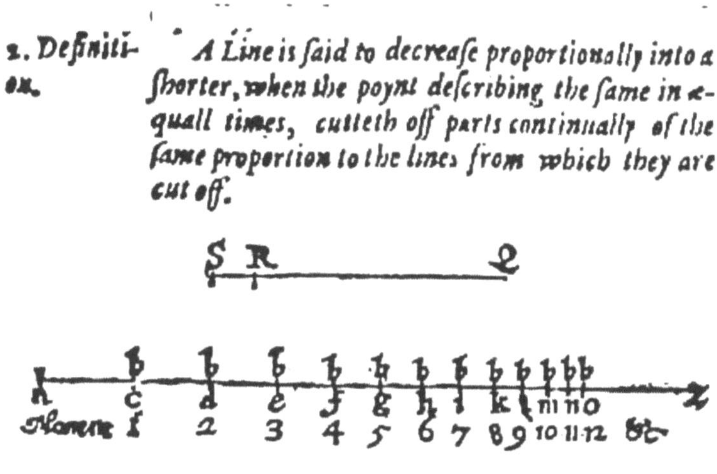

Figure 4.[41]

The whole sine was represented by aZ and it could be diminished proportionally by the motion of a point. This point cut off proportional parts of the whole sine, and then of the subsequent remainder in equal times. This proportion can be represented by

$$QR/QS = cZ/aZ = dZ/cZ = \ldots \qquad (4)$$

Napier here cited Euclid's *Elements* VI 10.[42] Next, he presented a corollary, "Hence it followeth that by this decrease in equall moments (or times) there must needes also bee left proportionall lines of the same proportion."[43] Napier again referred to the *Elements*, this time V 19 and VII 11.[44] Care is taken to support the corollary for both numbers and geometric magnitude. If we let the whole sine, or aZ, be denoted R, $(1 - 1/R) = P$ and let $ac = 1$ unit. Then

$$cZ = aZ - ac = R - 1 = R(1-1/R) = RP \qquad (5)$$

$$dZ = cZP = RP^2 \qquad (6)$$

$$eZ = dZP = RP^3 \qquad (7)$$

[41] Nepair, *A Description*, p.2.
[42] Euclid, *Elements*, trans. T. Heath, vol.II. (New York: Dover, 1956), p.213. Proposition 10 reads, "To cut a given uncut straight line similarly to a given cut straight line."
[43] Nepair, *A Description*, p.3.
[44] Euclid, *Elements*, V 19 "If, as a whole is to a whole, so is a part subtracted to a part subtracted, the remainder will also be to the remainder as whole to whole., p. 174 and VII, 11, "If, as whole is to whole, so is a number subtracted to a number subtracted, the remainder will also be to the remainder as whole to whole," p.311.

and so forth. Napier, of course, phrased these ideas in the language of proportions and motion.[45]

Napier went on to discuss the representation of irrational magnitudes in his third definition. He stated that although "surd" quantities were "inexplicable" by number that they could be considered to be approximated closely enough by numbers that differed by less than one "from the true value of the Surd quantities."[46] He needed to consider the representation of irrational quantities by numbers so that later he could give numerical answers for logarithms that were the measure of irrational quantities. Indeed, it seems that Napier knew that every point on a line could be given a numerical approximation. We see numbers being associated with continuous magnitude rather than discrete units.

Napier continued with definition four, stating that "Equall-timed motions are those which are made together, and in the same time."[47] He then considered the two lines *Ai* and *aZ*, letting movement from *A* to *C* on the first line take place at the same time as motion from *a* to *c* on the second line. Therefore *AC* and *ac* are equal-timed motions. Next, definition five is "Seeing that there may bee a slower and a swifter motion given then any motion, it shall necessarily follow, that there may be a motion given of equall swiftnesse to any motion."[48] And lastly, Napier completed his definition of logarithms. Napier stated in definition six:

> The Logarithme therefore of any sine is a number very neerely expressing the line, which increased equally in the meane time, whiles the line of the whole sine decreased proportionally into that sine, both motions being equal-timed, and the beginning equally swift.[49]

These last two definitions are notable, as nobody had expressed anything in these terms before. A developed calculus would have given Napier the tools he needed to deal with a variable velocity, but he had no such methods available to him. Definition five seems only to be necessary as a sort of intuitive guarantee of continuity. Kepler, in his work on logarithms, thought there was no justification in logic for Napier's use of motion as an aid to calculation. Briggs, although he continued Napier's work, did not utilise Napier's kinematic foundation. However, as we will see later in the mathematics of Barrow, kinematic techniques were later employed for different sorts of problems in English mathematics.

It has been pointed out several times elsewhere that strictly speaking because the velocity of the moving point on the geometric line varies continually, its proper expression calls for the calculus.[50] Therefore, as Napier could not deal with continuous

---

[45] "let the whole sine aZ be diminished proportionately - let the poynt diminishing the same by his motion be b, and the proportion of each part to the line from wch it is cut off, be as QR to QS. Therefore in what proportion QS is cut in R. in the same proportion (by the 10 of the 6 of Euclid) Let aZ be cutt in c and so be b running from a to c in the first moment, cutt off ac from aZ, the line or sine cZ remaining."

[46] Napair, *A Description*, p. 4.

[47] Nepair, *A Description*, p.4.

[48] Nepair, *A Description*, p.4.

[49] Nepair, *A Description*, pp.4-5.

[50] See, for example, Morris Kline, *Mathematical Thought from Ancient to Modern Times*, p. 256 and Baron,

functions he calculated the bounds between which the logarithms must lie. He began this process in definition six. The logarithm of the whole sine, in this case 1,000,000, was said to be zero and consequently the logarithms of numbers greater than the whole sine were negative. This is so by definition, as the sines decrease from the whole sine the logarithms increase from zero. Napier went on in chapter two, "On the Propositions of Logarithms," to explore propositions that would allow one to actually calculate logarithms.

Napier framed all of his propositions in terms of proportions. Charles Naux, in his *Historie des Logarithms de Napier à Euler*, states

> Neper a composé sa théorie à l'aide des seules données du V livre d'Euclide, domptées par son génie et adaptées à une construction originale pour qui elles n'avaient pas été prévues.[51]

It does seem that Napier depended heavily upon Euclid's *Elements* Book V for his work with proportions of logarithms. Indeed, Euclid and Archimedes are the only mathematicians cited in Napier's works. Herman Goldstine remarks that the ease with which Napier handled velocities "shows quite clearly the degree of sophistication attained in the West by 1600 in coping with nonuniform motions."[52] He does not, however, expand on this remark or provide any examples to support it. Perhaps he had in mind, for instance, Kepler's application of mathematics to a nonuniform motion in a noncircular orbit using infinitesimals.[53] It is still, however, difficult to ascertain what work might have led Napier to use points moving in space in the way he did in his construction of logarithms. Archimedes in his work, *On Spirals*, used points moving at uniform rates, but he did not employ the concept of nonuniform acceleration,[54] nor did he directly associate magnitude and number. The Merton School at Oxford had also used kinematic representations as early as the 14th century in discussion of concepts such as speed.[55] They developed mathematical descriptions of categories of motion such as uniform and constantly accelerated motion and considered the possible relations between impressed force and the velocity produced, but their work was geometric. Nicole Oresme introduced diagrams that pictured the effect of an active quality. For instance, the Mertonian rule stated that in a motion uniformly accelerated from rest to a velocity v, the total distance traversed is the same as that covered in the same time by a motion at a constant velocity $v / 2$. This was depicted pictorially using triangles and rectangles to represent the distance covered and the motions elapsed over the course of

---

DSB, p. 612.

[51] Charles Naux, *Historie des Logarithmes de Nepier à Euler* (Paris: A. Blanchard, 1966), p.13

[52] Herman Goldstine, *A History of Numerical Analysis from the 16th through the 19th Century* (New York: Springer-Verlag, 1977), p.6.

[53] For details concerning this example, see Eric J. Aiton, "Infinitesimals and the Area Law," *Internationales Kepler - Symposium Weil der Stadt 1971*, ed. F. Krafft, K. Meyer, and B. Sticker (Hildesheim: Dr. H.. . Gerstenberg, 1973), pp.285-305.

[54] Archimedes, *The Works of Archimedes*, ed. T.L. Heath (New York: Dover, 1953), pp.155. He did, however, in propositions 10 and 11 use *n* lines that formed an ascending arithmetic progression.

[55] D.T. Whiteside, "Mathematical Thought in the Later 17th Century," *AHES*, Vol. 1., p.215.

time.[56] This earlier work, however, did not associate numbers with geometrical magnitudes. Napier was one of the first to use geometrical magnitude as a foundation for a broadened number concept. He, however, remained inside the classical tradition in his use of proportion.

Chapter two can be quickly paraphrased. If we have proportional numbers, $a/b = c/d$ then

$$\log a - \log b = \log c - \log d \tag{8}$$

$$(b = c)\ \ 2\log b - \log a = \log d \tag{9}$$

$$(b = c)\ \ 2\log b = \log a + \log d \tag{10}$$

$$(b>c)\ \ \log b + \log c - \log a = \log d \tag{11}$$

$$(b< c)\ \ \log b + \log c = \log a + \log d \tag{12}$$

also, if $a/b = b/c = c/d$

$$3\log b = \log d + 2\log a \quad \text{and} \quad 3\log c = \log a + 2\log d. \tag{13}$$

Napier himself, however, put these propositions in terms of Euclidean proportionals, for example, proposition eleven was stated "Of the Logarithms of four proportionals, the summe of the second and third, made lesse by the first, is equal to the fourth."[57] Napier did not supply more information as to the making of logarithm tables. He merely provided the tables and the above propositions that would help one to find logarithms of sines not included in the tables. He worried about how the logarithms would be received and wanted to wait to publish his method of construction until after he tested the waters because he "expect[ed] the judgement and the censure of learned men hereupon, before the rest rashly published, be exposed to the detraction of the envious."[58]

---

[56] Olaf Pedersen, *Early Physics and Astronomy: A Historical Introduction* (Cambridge: Cambridge University Press, 1974), pp.191-213.
[57] Nepair, *A Description*, p.8.
[58] Nepair, *A Description*, p.10.

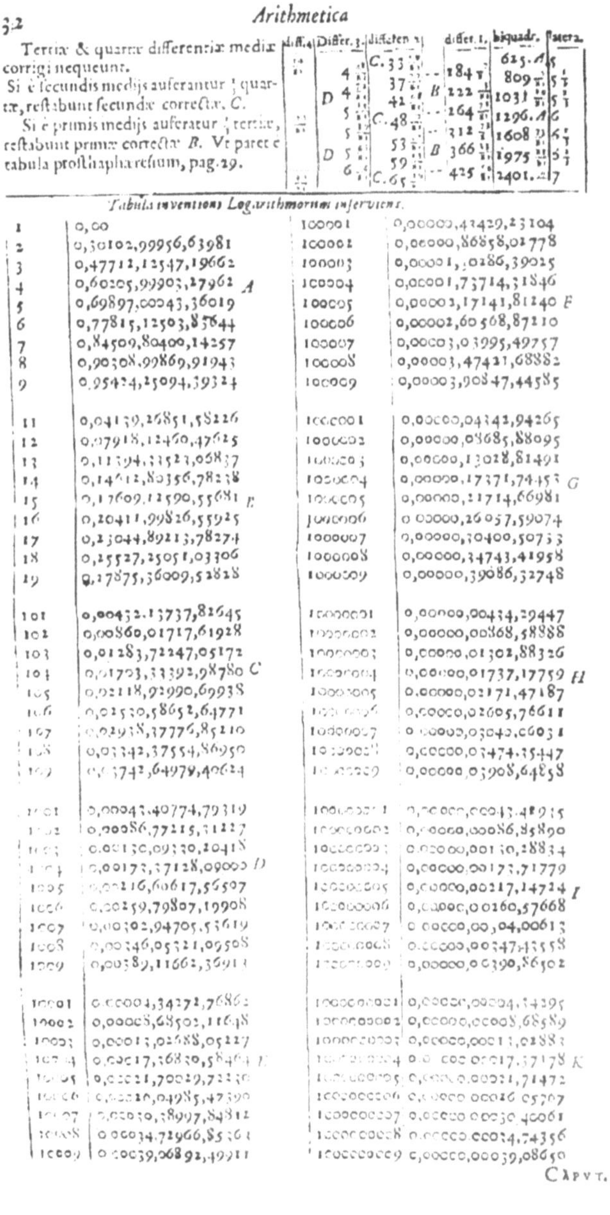

32 *Arithmetica*

Tertiæ & quartæ differentiæ mediæ corrigi nequeunt.
Si è secundis medijs auferantur ¼ quartæ, restabunt secundæ correctæ. C.
Si è primis medijs auferatur ¼ tertiæ, restabunt primæ correctæ B. Vt patet è tabula prosthaphæresium, pag. 29.

diff. 4	Differ. 3	differen. 2		differ. 1	biquadr.	latera
		C. 33			625. A	5
	4			184	809	5
		37	B	222	1031	5
	D 4	42				
	5	C. 48		264	1296. A	6
	5			312	1608	6
	D 5	53	B	366	1975	6
		59				
	6	C. 65		425	2401. A	7

Tabula inventioni Logarithmorum inserviens.

1	0,00	100001	0,00000,43429,23104
2	0,30102,99956,63981	100002	0,00000,86858,01778
3	0,47712,12547,19662	100003	0,00001,30286,39025
4	0,60205,99903,27962 A	100004	0,00001,73714,31846
5	0,69897,00043,36019	100005	0,00002,17141,81240 F
6	0,77815,12503,83644	100006	0,00002,60568,87210
7	0,84509,80400,14257	100007	0,00003,03995,49757
8	0,90308,99869,91943	100008	0,00003,47421,68882
9	0,95424,25094,39324	100009	0,00003,90847,44585
11	0,04139,26851,58226	1000001	0,00000,04342,94265
12	0,07918,12460,47625	1000002	0,00000,08685,88095
13	0,11394,33523,06837	1000003	0,00000,13028,81491
14	0,14612,80356,78238	1000004	0,00000,17371,74453 G
15	0,17609,12590,55681 B	1000005	0,00000,21714,66981
16	0,20411,99826,55925	1000006	0,00000,26057,59074
17	0,23044,89213,78274	1000007	0,00000,30400,50733
18	0,25527,25051,03306	1000008	0,00000,34743,41958
19	0,27875,36009,52828	1000009	0,00000,39086,32748
101	0,00432,13737,82645	10000001	0,00000,00434,29447
102	0,00860,01717,61928	10000002	0,00000,00868,58888
103	0,01283,72247,05172	10000003	0,00000,01302,88326
104	0,01703,33392,98780 C	10000004	0,00000,01737,17759 H
105	0,02118,92990,69938	10000005	0,00000,02171,47187
106	0,02530,58652,64771	10000006	0,00000,02605,76611
107	0,02938,37776,85210	10000007	0,00000,03040,06031
108	0,03342,37554,86950	10000008	0,00000,03474,35447
109	0,03742,64979,40624	10000009	0,00000,03908,64858
1001	0,00043,40774,79319	100000001	0,00000,00043,42915
1002	0,00086,77215,31227	100000002	0,00000,00086,85890
1003	0,00130,09330,20418	100000003	0,00000,00130,28834
1004	0,00173,37128,09000 D	100000004	0,00000,00173,71779
1005	0,00216,60617,56507	100000005	0,00000,00217,14724 I
1006	0,00259,79807,19908	100000006	0,00000,00260,57668
1007	0,00302,94705,53619	100000007	0,00000,00304,00613
1008	0,00346,05321,09508	100000008	0,00000,00347,43558
1009	0,00389,11662,36913	100000009	0,00000,00390,86502
10001	0,00004,34272,76862	1000000001	0,00000,00004,34295
10002	0,00008,68502,11648	1000000002	0,00000,00008,68589
10003	0,00013,02688,05227	1000000003	0,00000,00013,02883
10004	0,00017,36830,58464 E	1000000004	0,00000,00017,37178 K
10005	0,00021,70929,72230	1000000005	0,00000,00021,71472
10006	0,00026,04985,47390	1000000006	0,00000,00026,05767
10007	0,00030,38997,84812	1000000007	0,00000,00030,40061
10008	0,00034,72966,85361	1000000008	0,00000,00034,74356
10009	0,00039,06892,49911	1000000009	0,00000,00039,08650

CAPVT.

Figure 5. A table of Logarithms.

The third chapter described the table of logarithms and the contents of its columns. Napier, in the fourth chapter, gave examples of the uses of his table. He showed how to find the logarithms in the table of sines, tangents, and secants. Of course, not all values are included in the table, so Napier included some remarks on how to calculate logarithms that could not be found exactly in the table. He ends the first book with a fifth chapter that provides many examples of logarithm problems that utilised proportions. Napier concludes with the remark,

> Now out of this that is already delivered, let the learned judge how great the benefit the *Logarithms* bring them; seeing that by the addition and subtraction of them, and by dividing by *2* and *3*, and by other easie additions, or subtractions, multiplication, divisions: the extraction of the square and cubicke rootes, and all the great toyle of calculating is avoided...[59]

Napier had no cause for concern; the *Descriptio* was indeed a very well received book. In his doctorial thesis, W.F. Hawkins includes a chapter titled "The Explosion of Logarithms 1614-1640." He lists at least twenty-two books written on logarithms by 1633.[60] Napier's next publication was not as popular, as it dealt with the construction of logarithm tables in a fashion that was soon superseded.

Napier's *Mirifici Logarithmorum Canonis Constructio*, hereafter referred to as the *Constructio*, was first published in 1619 two years after his death. This work was only reprinted once, in 1620, and it was not translated into English until 1889.[61] Although it covers all the material discussed in the *Descriptio*, the *Constructio* includes the actual details of his construction of the tables and makes Napier's concern with accuracy of his approximations clearer. The *Constructo* was not as popular or as influential as the *Descriptio* for several reasons. Briggs had published by 1617 a table called *Logarithmorum Chilias Prima* with base ten logarithms that quickly superseded Napier's version. Moreover, the *Constructio's* kinematic foundations were replaced by Briggs' directly numerical techniques.

Napier, in the *Constructio*, labelled the line created by the uniform velocity of a point "arithmetical." The sine line created by nonuniform acceleration of a point was labelled "geometric." Before he explained how these concepts would help him to construct the actual numerical tables, he first introduced rules to insure the accuracy of his calculations within certain limits. He was especially concerned to show that one could represent irrational magnitudes numerically with acceptable accuracy. In this section he gave an example from Archimedes that showed he was familiar with Archimedes' methods of approximating irrationals. Napier then began to actually construct his tables.

Basically, Napier's goal was to obtain two limits between which a logarithm must lie, and his whole method of construction depended upon the use of these limits, together with corresponding limits for the value of the difference of the logarithms of

---

[59] Nepair, *A Description,* p.29.
[60] Hawkins, p.161.
[61] John Napier, *The Construction of the Wonderful Canon of Logarithms*, trans. William Macdonald (London: Dawsons of Pall Mall, 1966), p.xix.

two numbers. In order to assist his calculations, Napier discussed the construction of arithmetical and geometric progressions. He stated, "The construction of every arithmetical progression is easy; not so, however, of every geometrical progression."[62] The only geometrical progression he believed to be easily constructed was one that arose from the subtraction of "an easy part of the number from the whole number." To that end, he explained how to find the tenth, hundredth and thousandth parts of numbers. He now had all his tools in place to construct his first table.

The first table consists of 101 numbers, with 10^7 (the whole sine) being first, and $1-1/10^7$ being the common ratio. In modern notation, the table for $r = 0$ to 100 is:

$$10^7\,(1-1/10^7)^r \tag{14}$$

The second table contains 51 numbers $10^7(1 - 1/10^5)^r$, where $r = 0,1, \ldots 50$. Napier stated that he had picked "the proportion which is easiest, and as near as possible to that subsisting between the first and last numbers of the First table."[63] This is so because the common ratio $(1 - 1/10^5)$ is nearly equal to $(1-1/10^7)^{100}$, that of the last number in the first table, to the first in the last.

Unfortunately, Napier made an arithmetical error in calculating the last entry in this table, the number should be 9,995,001.224804. This caused an error in the last place in Napier's table of logarithms. The table of logarithms had other problems due to errors in the table of sines he used,[64] but when one considers the number of calculations that Napier made it is remarkable that he made only a few arithmetical errors.

The third table was much larger than the others. It contained 69 different columns, and each column has 21 numbers. In the second table, the ratio of the last to the first is $(1-1/10^5)^{50}$, which is close to $1-1/2000$. The first number in any column in the third table is obtained by taking $1-1/100$ of the first number in the preceding table. The numbers in any one column are obtained by successive multiplication by $1-1/2000$, so the pth number in the qth column is

$$10^7\,(1-1/2000)^{p-1}\,(1-1/100)^{q-1} \tag{15}$$

The first term in the first column was the whole sine, or 10^7 and the last term in column 69 was 4,998,609.4034. Thus the third table finally covered the interval from radius to half radius.[65]

[62] Napier, *The Construction*, p.11.
[63] Napier, *The Construction*, p.13.
[64] Goldstine, p. 7.
[65] Napier, *The Construction*, pp.14-16.

Table 1.

First Column	**Second Column**	**69th Column**
10000000.0000	9900000.0000	5048858.8900
9995000.0000	9895050.0000	5046334.4605
9990002.5000	9890102.4750	5043811.2932
9985007.4987	9885157.4237	5041289.3879
&.c.	&.c.	&.c.
Continuously	Descending	finally
To	To	to
9900473.5780	9801468.8423	4998609.4034

Each of the twenty-one numbers in the first table is used as the first term of a geometric progression with the common ratio *99/100*. He then could take his third table of sines, "or natural numbers decreasing geometrically," and fit them with "logarithms or artificial numbers increasing arithmetically."[66]

Napier was now ready to define his logarithms. He stated,

The logarithm of a given sine is that number which has increased arithmetically with the same velocity throughout as that with which radius began to decrease geometrically, and in the same time as radius has decreased to the given sine.[67]

Napier let TS be the radius of the whole sine, and let *ds* be a given sine in the same line, letting *g* move geometrically from T to *d* in certain, determinate moments of time. Then, he let *bi* be another line, infinite towards *i*, along which, from *b*, he let *a* move arithmetically with the same velocity which g had at first when at T. From the fixed point *b* in the direction *i* let a advance in just the same moments of time up until the point *c*. Then the number measuring the line *bc* is called the logarithm of the given sine *ds*.

[66] Napier, *The Construction*, pp.16-17.
[67] Napier, *The Construction*, p.19.

26. *The logarithm of a given sine is that number which has increased arithmetically with the same velocity throughout as that with which radius began to decrease geometrically, and in the same time as radius has decreased to the given sine.*

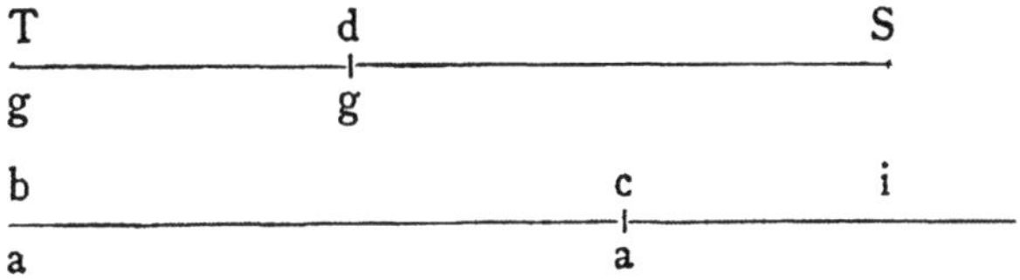

Let the line T S be radius, and d S a given sine in the same line; let g move geometrically from T to d in certain determinate moments of time. Again, let b i be another line, infinite towards i, along which, from b, let a move arithmetically with the same velocity as g had at first when at T; and from the fixed point b in the direction of i let a advance in just the same moments of time up to the point c. The number measuring the line b c is called the logarithm of the given sine d S.

27. *Whence nothing is the logarithm of radius.*

For, referring to the figure, when g is at T making its distance from S radius, the arithmetical point d beginning at b has never proceeded thence. Whence by the definition of distance nothing will be the logarithm of radius.

Figure 6.

In this manner Napier effectively built up a one-to-one correspondence between the points on the arithmetic and geometrical lines. He provided extraordinary numerical approximations of what we would now consider to be a continuous function. The continuity of the 'function' was founded on the continuously moving points. Napier depended on notions of time, motion and instantaneous speed. Of course, he did not have our formal notion of function. But if we conceive of a function, not as a formula, but as a general relation associating the elements of one set of numbers, in this case the sines, with the numbers of another set, in this case the logarithms, Napier did have the notion of function.

Now that Napier had his relationships in place, he could begin to calculate his logarithms. As in the *Descriptio*, he began by noting that the logarithm of the whole sine, or radius, is zero by the definition of distance.[68] In order to continue his computations he introduced the concept of the limits of the logarithms.

Whence also it follows that the logarithm of any given sine is greater than the difference

[68] Napier, *The Construction*, p.19.

> between radius and the given sine, and less than the difference between radius and the quantity which exceeds it in the ratio of radius to the given sine. And these differences are therefore called the limits of the logarithms.[69]

He based his argument in terms of the distance covered by points moving at different velocities.

First, in figure 5, Napier produced ST beyond T to *0*, so 0S is to TS as TS to dS. Then *bc*, the logarithm of the sine dS, is greater than Td and less than 0T. Napier reasons that this must be so because in the same time that g moves from T to *d* because by his definition of geometrical movements, 0T is to 0S as Td is to TS, and in the same time (by definition of logarithms) is *a* moved from *b* to *c*, so that oT, Td and *bc* are distances travelled in equal times. Moreover, g when moving between *T* and *0* is swifter than at T, and between T and *d* slower, but at T is equally swift with *a* . (by definition of logarithms). It follows then that 0T (the distance travelled by g moving swiftly) is greater, and Td (the distance travelled when g is moving slowly) is less, than *bc*, the distance travelled by *a* with its "medium motion" in just the same moments of time. Therefore, he claimed that 0T is the greater limit and Td the lesser limit of the logarithm which *bc* represents. These limits were then used to find the logarithm of a given sine.[70] For example, in the first table, the logarithm of 9,999,999, is by the above notion of limits, between 1.000,000,1 and 1.000,000,0, so Napier set the arithmetic mean 1.000,000,5 for the required logarithm.

[69] Napier, *The Construction*, p.20.
[70] Napier, *The Construction*, p.20.

20 CONSTRUCTION OF THE CANON.

28. *Whence also it follows that the logarithm of any given sine is greater than the difference between radius and the given sine, and less than the difference between radius and the quantity which exceeds it in the ratio of radius to the given sine. And these differences are therefore called the limits of the logarithm.*

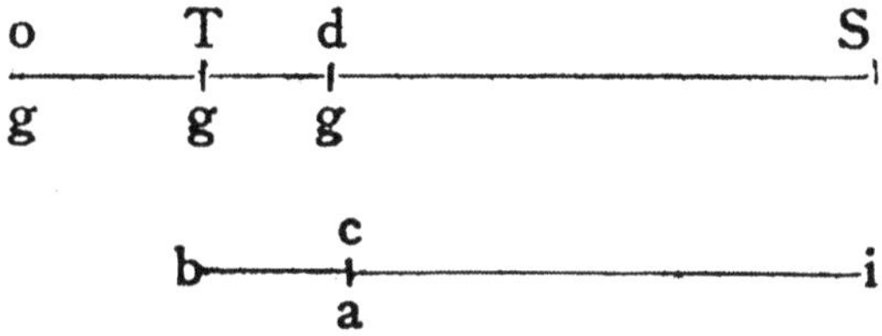

Thus, the preceding figure being repeated, and S T being produced beyond T to o, so that o S is to T S as T S to d S. I say that b c, the logarithm of the sine d S, is greater than T d and less than o T. For in the same time that g is borne from o to T, g is borne from T to d, because (by 24) o T is such a part of o S as T d is of T S, and in the same time (by the definition of a logarithm) is a borne from b to c; so that o T, T d, and b c are distances traversed in equal times. But since g when moving between T and o is swifter than at T, and between T and d slower, but at T is equally swift with a (by 26); it follows that o T the distance traversed by g moving swiftly is greater, and T d the distance traversed by g moving slowly is less, than b c the distance traversed by the point a with its medium motion, in just the same moments of time; the latter is, consequently, a certain mean between the two former.

Therefore o T is called the greater limit, and T d

Figure 7.

Napier then followed the same organisation of material he had in the *Descriptio*, giving rules for the operation of logarithms. Once again, these rules were all stated in the language of proportion. He then provided many actual numerical examples that showed how to work with the tables and the rules.

Another way to express Napier's idea is to assume a particle begins at A and moves toward Z with a speed at each point proportional to the remaining distance to Z with a speed at each point proportional to the remaining distance to Z. Thus if the speed of the particle at B is v_b we have

$$V_b = k(BZ). \tag{16}$$

Suppose the constant $k = 1$ so that

$$V_b = (BZ). \tag{17}$$

Let the initial speed be g. Hence

$$g = AZ. \tag{18}$$

Take now a small time period Δt. Let $B, C, D, E,$ .... be the points occupied by the particle at times Δt, $2(\Delta t)$, $3(\Delta t)$, . . .

A B C D Z

Assume for the sake of analysis that the speed within each time interval Δt is constant. Thus the speed of the particle along AB is $g = AZ$; along BC, BZ; along CD, CZ and so on. Let

$$\alpha = (1 - \Delta t) \tag{19}$$

Then we have

$$AZ = (AZ)\alpha^0 = g\alpha^0 \tag{20}$$

$$BZ = AZ - AB = AZ - g\Delta t = g - g\Delta t \tag{21}$$

$$= g(1 - \Delta t) \tag{22}$$

$$= g\alpha \tag{23}$$

$$CZ = BZ - BC = g\alpha - (g\alpha)\Delta t = g\alpha(1 - \Delta t) \tag{24}$$

$$= g\alpha^2 \tag{25}$$

$$DZ = CZ - CD = g\alpha^2 - (g\alpha^2)\Delta t = g\alpha^2(1 - \Delta t) \tag{26}$$

$$= g\alpha^3 \tag{27}$$

and so on.

Suppose now that a second particle moves along a line $A'B'$ with a **constant** speed equal to the initial speed g of the first particle.

A' B' C' D' E' Z'

Let B', C', D', . . . be the points occupied by the second particle at times Δt, $2(\Delta t)$, $3(\Delta t)$, . .
We have

$$A'B' = (\Delta t)g \tag{28}$$

$$A'C' = 2(\Delta t)g \tag{29}$$

$$A'D' = 3(\Delta t)g \tag{30}$$

$$A'E' = 4(\Delta t)g \tag{31}$$

And so on.

We can now form a table correlating the distances AZ, BZ, CZ, DZ, ... of the first particle to the distances $A'B'$, $A'C'$, $A'D'$, $A'E'$, . . . of the second particle:

Table 2.

$AZ = g\,\alpha^0$	$BZ = g\alpha$	$CZ = g\alpha^2$	$DZ = g\alpha^3$	Geometric progression
0	$A'B' = (\Delta t)g$	$A'C' = 2(\Delta t)g$	$A'D' = 3(\Delta t)g$	Arithmetic progression

Given x, assume $x = (AZ)\alpha^n = g\alpha^n$ for some n. The Napier logarithm of x is equal to $n(\Delta t)g$.[71]

As is shown in the above discussion, Napier did not construct his logarithms in terms of bases, in the modern sense of the word, although his system can be thought of as logarithms to the base $1/e$. The now familiar rules for the logarithms of products, quotients, and exponents did not hold because he chose the whole sine rather than one as the logarithm whose base was zero.[72] His discussions with Henry Briggs in 1615 and 1616 led to the idea of changing the system so that zero would become the logarithm

[71] I would like to thank my supervisor, Dr. Craig Fraser, for introducing me to this way of expressing Napier's ideas.
[72] Baron, *DSB*, p. 612.

of one and 10^{10} that of the whole sine. Briggs claimed that Napier had wanted to make this change before the publication of the *Descriptio*, but had not been able to perform the calculations due to his failing health. Thus it was left to Briggs to reformulate the tables by adapting them to a decimal base.

BRIGGS AND NUMERICAL METHODS

Henry Briggs (1561 - 1630) went to St. John's College, Cambridge where he received his B.A. in 1581 and his M.A. in 1585. He began his teaching career as an examiner and lecturer in mathematics in 1592 after being elected a fellow of his college in 1589. Briggs became the first professor of geometry at Gresham College in London in 1596, and he was also the first Savilian professor of geometry at Oxford from 1619 to his death in 1630.

Although Briggs was a leader in English mathematical education, it is more likely his interest in navigation that led him to devote himself to logarithms. He was a member of William Gilbert's circle at Wingfield house, and at Gilbert's request he calculated a table of magnetic dip and variation. He also recorded and tabulated much of the information used by Edward Wright in his 1599 *Certain Errours in Navigation detected.* His interest in navigation was such that he acted as a link between Gresham College and a group of master shipwrights centred around John Wells, Keeper of Naval Stores at Deptford. Briggs helped many seamen with mathematics, and he seems to have inspired the respect of those he tutored; for example, a sea-captain named some north-western islands he discovered "Briggs his Mathematickes."[73] Navigators enthusiastically adopted logarithms as soon as they were developed and as soon as Briggs learned of the logarithms he began to lecture on them at Gresham College.[74]

Briggs's interest in navigation is also shown by his connection to the Virginia Company. Briggs's earliest connection with the Virginia Company was as a consultation about the Northwest passage due to the obscurity of the information brought back by Thomas Button after his westward sweep of Hudson's Bay.[75] Briggs had also been involved in Button's voyage, insofar as he was one of the original 288 members of the Northwest passage company, which was incorporated under the personal patronage of Prince Henry.[76] Apparently, Briggs's qualifications as a mathematician were supposed to be useful in weighing the evidence of tides and currents observed by Button.

---

[73] Christopher Hill, *The Intellectual Origins of the English Revolution* (Oxford: Clarendon Press, 1965), pp. 37-51.

[74] Soon most work on navigation, like Matthew Norwood's *System of Navigation,* published in 1692, included a section discussing logarithms.

[75] Button's voyage was clouded by secrecy. Perhaps because of the combination of Prince Henry's patronage and the Princes' early death before Button's return, information about this voyage was never published. Purchas was unable to obtain an account for his book, and Brigg's himself was "kept, to a certain degree, in the dark." It was felt, however, that discovery of the passage was imminent. See, *Narratives of Voyages towards the North-West in Search of a Passage to Cathay and India 1469 to 1631* (ed) Thomas Rundall, esq. (London: Hakluyt Society, 1869), p. 81.

[76] Andrews, *Trade, Plunder and Settlement*, p. 349.

Briggs's greater involvement with the Virginia Company, due to his election as an auditor of accounts on 28 April, 1619, increased his participation in exploration, colonisation and navigation. There were two major contributing factors. First Briggs took sides in a faction fight concerning the control and aims of the Virginia Company. Second, an "Indian massacre" threatened the colony's very existence. Together, these difficulties stimulated Briggs to make efforts to save the colony. In this endeavour, he utilised his mathematical reputation in several ways. First, he used his reputation as a mathematician to lend weight to a series of company audits. Next, he drew up maps showing the existence of a Northwest passage and encouraged the notion of Virginia as an important jumping off point for further exploration.

It was, however, the so-called "Indian massacre" of 1622, that increased Briggs's direct involvement in Northwest passage exploration. In March of 1622, the natives attacked the widely scattered and poorly defended groups of planters in Virginia. Over three hundred men, women and children were killed, and the colonists had to abandon over half of their settlements. Disastrously, as the company itself was almost bankrupt, the colony could expect little by way of future supply. In this atmosphere, many wondered if the colony should be evacuated.[77] Even before the "massacre", there was disappointment about the lack of quick profit from fur trading or precious metals. Many of the City merchants did not have the patience to wait and develop plantation products; they sought an immediate return.[78] The task, for those supporting the colony's future existence, was to present Virginia as a great undertaking "full of hope and worthy of the support of all men."[79] Worried about Virginia's reputation, Edward Waterhouse wrote *A Declaration of the State of Virginia*, and he included Briggs's "Treatise of the Northwest Passage to the South Sea, through the Continent of Virginia." Waterhouse held out the possibility of "a rich trade to Cathay, China & Japan" through Virginia, using Briggs's tract to support the continuance of the colony both as a staging ground for future ventures, and as possible Northwest passage route. The authority to pronounce on the feasibility of the Northwest passage routes seems to have rested in the description of Briggs as a "learned and famous Mathematician."[80]

When Briggs wrote about Northwest passage issues, his account was certainly reassuring. A passage to the south seas could be found in the west and north-west part of Virginia; he wrote that "on the other side of the mountains and beyond the falls . . . [there lies] a free and fair passage to China, Japan, New Spain, Peru and Chila."[81] One might wonder, however, why had such a passage not yet been found? The difficulty rested in the "crafty falsehood of these usual Maps"; Briggs implied that Spanish and Portuguese maps were purposefully misleading "to discourage such as otherwise might

---

77. Craven, *Dissolution of the Virginia Company*, p. 195-208.
78. R. Brenner, "The social basis of English commercial expansion 1550-1650" *Journal of Economic History* 32(1972), 361-841, p. 374-377.
79. Craven, *Dissolution of the Virginia Company*, p. 208.
80. Edward Waterhouse, *A Declaration of the State of Virginia . . .and a Treatise Annexed, Written by that learned Mathematician Mr Henry Briggs, of the Northwest passage to the South Sea through the Continent of Virginia . . .* (London, 1622).
81. Briggs inWaterhouse, *A Declaration of the State of Virginia*, p. 49.

bee desirous to search a passage by that way aforesaid into those Seas."[82] An incontrovertible proof was supplied for the passage's existence: there must be a passage because God wanted the heathen converted and the Plantation of Virginia to be a success.[83] Religious motivations were very important to Briggs. Lilly tells us that he was a "severe Presbyterian", and Briggs was active in causes such as trying to raise funds for the relief of Calvinist academics who had fled from the Platinate.[84] More generally, even from the first voyage of Martin Frobisher, in 1576, the trope of the "conversion of the heathen" became increasingly prominent in discussions of westward expansion.[85] For example John Davis's *The Seaman's Secrets* (London, 1594) claimed that navigation was a divine science because it made possible the spread of the word of God, and helped Protestant Christianity drive back Catholic Spain.

Briggs's involvement with the Northwest passage did not end with his literary contributions. Along with his expert advice, he also helped to procure capital for a voyage. He arranged for Sir John Brooke and "divers of their friends" to support Captain Luke Foxe's attempt to find the Northwest passage. Captain James, of Bristol, also attempted to procure Briggs's support for his venture; he went to visit Briggs in Oxford just before Briggs's death.[86] Briggs was viewed as a potentially valuable patron to James; he hoped with Briggs's support to be put in charge of Foxe's ship, wanting to be the admiral of a joint venture. Without Briggs's influence, James had to rely on local merchants, instead of the court, to fiance his journey. In some respects, with regard to Foxe's venture, Briggs himself must be seen as a low-level patron. At the very least, he used his contacts and his reputation to further Foxe's journey. Even in death, as earlier noted, Briggs showed his devotion to north-west passage exploration by leaving ten pounds towards its discovery in his will.[87] Unfortunately, Briggs died before either man could set out, but both parties used his map.[88] Briggs's interest in navigation coupled with his connections to Northwest passage voyages indicate that his development of logarithms was partially in aid of practical pursuits.

BRIGGS' METHOD OF CONSTRUCTION

It is difficult to directly calculate the logarithm of a prime number, and Briggs

---

82. Briggs, inWaterhouse, *A Declaration of the State of Virginia*, p. 49.

83 Briggs's manuscript material also shows that he believed his north-west passage work was undertaken for the glory of God. See, for example, Bodleian MS 313 f. 65.

84 BL MS Tanner 72 fol. 211 and fol. 228.

85. Louis B. Wright, *Religion and Empire: The Alliance between Piety and Commerce in English Expansion 1558-1625* (Chapel Hill: University of North Carolina Press, 1943) argues that prospects of conversion became important as a motive, and given Briggs's history his desire for conversion was probably sincere. However, it should be recognised that this idea, along with ploys such as employment for the poor, were part of a standard package intended to show potential patrons the beneficial returns of a project. See also Andrews's *Trade, Plunder and Settlément*, p. 33

86. Luke Foxe, *North-West Fox, or Fox from the North-west passage*... (London, 1635),p. 170-171.

87 PROB 11/159 fol. (and p. 199) St John – I would like to thank Frances Willmoth for sending me a transcript of Briggs's will.

88. Hallowes, 'Henry Briggs, Mathematician', p. 88.

proceeded in steps. His work demonstrates great accuracy in taking very many successive square roots keeping up to thirty-two decimal places. First he developed a system of logarithms of base 10 by taking successive square roots of ten and placing them in a table. It consisted of fifty-four lines with log 10 = 1 and log 1 = 0. He let A_n be the number on the first line and a_n its logarithm. Then,

Table 3.

	A_n	Log A_n	a_n	
$(A_n)^{1/2}$	A_{n+1}	Log A_{n+1}	a_{n+1}	$a_n/2$
$(A_{n+1})^{1/2}$	A_{n+2}	Log A_{n+2}	a_{n+2}	$a_{n+1}/2$

or,

Table 4.

	Numbers	Logarithms
	10,000	1
1	3,1622 77660 16387 93319 98893 54	0,5
2	1, 7782 79410 03892 28011 97304 13	0,25
3	1,3335 21432 16322 40256 65389 308	0,125
	. . . .	. . .
53	1,0000 00000 00000 02556 39298 64006 470	0,00000 00000 00001 1102 23024
54	1,0000 00000 00000 01278 19149 32003 235	0,00000 00000 00000 05561 11 51
	1,0000 00000 00000 01	0,000000000000000004342 9448

Briggs, much like Napier, worked almost entirely without symbols. For example, he noticed that

$$(1 + x)^{1/2} = 1 + x/2 \tag{32}$$

provided x is small, but he stated this as "When the decimal part of a number begins with 'n' zeros, then to that degree of accuracy the square root of the number may be obtained by halving the mantissa."[89] A belief in the continuity of numbers allowed Briggs to interpolate, suggested by his statement that in the two columns "all the numbers which could exist are to be written down in increasing order" and that they filled in all the gaps in the calculated numbers.[90] It was also observed that the logarithms of numbers with the form $1 + x$ with x very small are to all practical purposes proportional to x.

These various facts were used together to directly calculate the logarithms of given

[89] Henry Briggs, *Arithmetica Logarithmica*, the table is given on page 10, and the discussion of this relation in chapter VI, pages 9 - 12. The translation given is not exact, Briggs did not use the symbol *n*.
[90] Hawkins, p.217, has an interesting discussion of this point. I have used his translation.

prime numbers. However, Briggs's calculations have been discussed in great detail by Herman Goldstine in his *A History Of Numerical Analysis from the 16th through the 19th Century* and by Derek Whiteside in his "Patterns of Mathematical Thought in the later Seventeenth Century."[91] Both of these authors have also explored how Briggs's development of the binomial theorem for $n = 1/2$ was related to his need to find simpler methods to extract successive square roots. As Briggs viewed this method only as a computing convenience, he did not notice its general significance. This may be due to his inadequate notation for powers.[92] As this material has already been explored in the secondary literature, I will only summarise Briggs's results and then I will discuss the 1631 posthumous English translation of his work titled, *Logarithmicall Arithmetike*. The 1631 work is not an exact translation of his earlier work; it leaves out theoretical details in order to give a wide variety of practical examples that show the usefulness of logarithms. The many ways that logarithms could be applied to ease calculation might partially explain the general lack of concern about the foundations of the decimal expansions. Briggs himself pointed out that it was easy to check one's calculations, so why worry overmuch about the ontological status of numbers that are so very useful?

Briefly, Briggs discovered that he could take successive square roots of a number p until it was expressible in the form

$$\log_{10}(1 + x) = x \log_{10}e = k'x \tag{33}$$

where

$$k' = \log_{10} e = 0.4342944810325182765. \tag{34}$$

Then by

$$\log_{10} p = 2^n \log_{10}(1 + x) = 2^n k'x, \tag{35}$$

one can immediately calculate the desired logarithm.[93] It is clear that

$$\log_{10} A = 0.43429\ldots(\log_e A). \tag{36}$$

Briggs must have discovered this relationship through inspection of numerical patterns in his efforts to establish a satisfactory method of computing the base 10 logarithms as easily as possible. Indeed, it was Briggs' discouragement concerning the enormous fatigue of calculations that led him to develop his method of differences and his

---

[91] In Goldstine, see chapter one section three; also see Derek Whiteside, "Patterns of Mathematical Thought in the later Seventeenth Century," *AHES*, Vol. 1, 1960.

[92] Whiteside points out that Briggs used the Bombelli ring notation, the same notation used by Stevin, for his free variable. Unfortunately the variable performs the triple function of distinguishing power of the same variable, different variables, and place-value in the decimal expansion. p.196.

[93] Goldstine, p. 15.

binomial expansion for $n = 1/2$. There seems, however, to have been a very small audience for details on the actual construction of logarithms, although there was a greater interest in learning how to apply the numbers given in the tables.

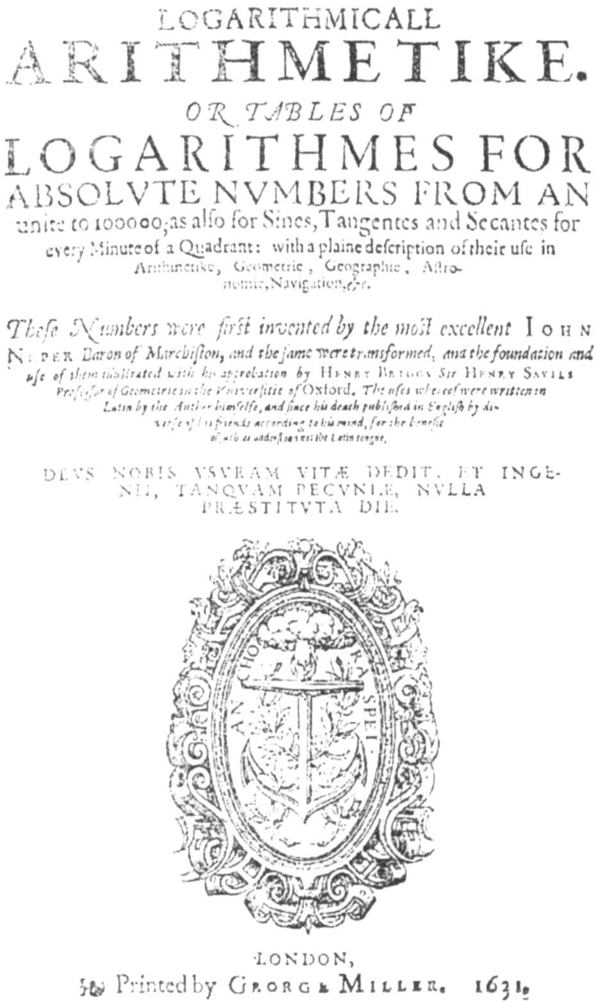

LOGARITHMICALL
ARITHMETIKE.
OR TABLES OF
LOGARITHMES FOR
ABSOLVTE NVMBERS FROM AN
unite to 100000; as also for Sines, Tangentes and Secantes for
every Minute of a Quadrant: with a plaine description of their use in
Arithmetike, Geometrie, Geographie, Astro-
nomie, Navigation, &c.

These Numbers were first invented by the most excellent IOHN NEPER *Baron of Marchiston, and the same were transformed, and the foundation and use of them illustrated with his approbation by* HENRY BRIGGS *Sir* HENRY SAVILS *Professor of Geometrie in the Vniversitie of* Oxford. *The uses whereof were written in Latin by the Author himselfe, and since his death published in English by diverse of his friends according to his mind, for the benefit of such as understand not the Latin tongue.*

DEVS NOBIS VSVRAM VITÆ DEDIT, ET INGENII, TANQVAM PECVNIÆ, NVLLA PRÆSTITVTA DIE.

LONDON,
Printed by GEORGE MILLER. 1631.

Figure 8. The Title page of Logarithmicall Arithmetike

The title page of *Logarithmicall Arithmetike* states that "The uses [of logarithms] whereof were written in Latin by the Author himselfe, and since his death published in English by diverse of his friends according to his mind, for the benefit of such as understand not the Latin tongue." This shows that there was an audience for logarithms among people who had less advanced educations. Moreover, all the theoretical details of construction were left out of this text, replaced by many practical examples of the use of logarithms. The book begins with instructions explaining how to look up the logarithm of a particular number in the tables. The construction of the logarithms was taken for granted.

The following chapters showed how easy multiplication and division of large numbers could be with the use of logarithms. More complicated examples such as the extraction of roots, the golden rule, and mean proportionals followed. But by page eleven, the examples turned to questions of interest to merchants. The first case study asked "A certaine summe of money forborne a cetaine time after it be due; what may it be worth at the end of the time according to a certaine rate propounded."[94] This problem is interesting for several reasons. First, it shows how quickly logarithms were adapted to financial calculations. Second, it shows logarithms being applied to time. The student is instructed to add the logarithm of the principal to the logarithm of the time, the sum being the logarithm of the principal and the increase, thus providing the true value sought. For example, suppose 375 pounds was loaned for 7 years with 8% interest,

2,57403,12677 - the Log of 375
0,2339,662885 - the difference of the Log of 100 and 108 multiplied by 7.
2,80799,75562 - the Log of 642 684/1000 or 642 pound 13 sh. 8 1/4 pence.

The text advised the student to have the differences between time periods such as years, months, weeks, and days always in readiness, and gives the student a list of handy numbers to reduce even further the need for multiplication and division. It also discussed how to apply logarithms for the calculation of leases, annuities, and pensions. It appears that it did not take long for people to appreciate that logarithms could ease calculations for many types of activities other than astronomy.

Chapter eight dealt with geometry problems that have no obvious real world applications. These examples show how logarithms can also simplify geometrical calculations. The first propositions are fairly simple, for example, "Having the two sides of a Parallelogramme; to find the Area thereof." The student is instructed to add the logarithms of the two given sides, and the sum will be the logarithm of the area required. Specific numbers are used to provide the student with a concrete example, if the student wanted to know the area of a parallelogram whose length is 50 and whose breath is 25, then

[94] Briggs, *Logarithmicall Arithmetike*, p.11

1,69897 - the Log. of 50
1,39794 - the Log. of 25
3,09691 - the sum, which is the Log. of 1250 the area required.[95]

The later propositions become more complex, and some of them relate to spherical geometry.

The next chapter is titled "Of the Resolution of Right-line Triangles." These problems require the student to use both trigonometric and logarithmic tables. For example, the second proposition is "Having two sides, and an Angle opposite to one of them, to find the other two Angles and the third side."[96] One interesting point about the solution is that it was given as a recipe–no justification of the steps taken was offered. The student is merely told to add the sine of the given angle to the log of the side opposite the angle required, and from this sum to subtract the log of the side opposite the given angle, the remainder was then declared to be the sine of the angle opposite to the other given side. A numerical example is given to ensure the clarity of the operations. For a triangle *ABC*, the side *AB* being 800, *BC* 320, and the angle *ABC* 128 grad. 4 min., the student was asked to find the angles *BAC, ACB,* and the side *AB*.

9,89613,68 - the sine of 128 gr. 4 min.
2,50515,00 - the log. of 320
12,40128,68 - the sum
2,90309,00 - the log of 800
9,49819,68 - the sine of 18 grad. 21 min for *BAC*

The proposition finishes by explaining that having *BAC* and *ABC*, the angle *ACB* is their complement to 180 grad., or 33 grad. 35 min. and that the side *AB* could be found by means of a previous proposition.[97] Perhaps the readers were considered to have a previous knowledge of the geometric material, as the careful explanations that were supplied to help the students set up the mercantile problems are not included.

In the sections concerning spherical geometry the propositions are also given like recipes without proofs or explanations. Examples, however, are provided that show how to find almost every unknown that might be sought. Perhaps the students were meant to use these propositions as templates for solving whatever problem they were actually interested in. Logarithms were shown to ease every type of calculation. The emphasis was on the problems themselves, and the foundation of the logarithms and the status of the numbers used was never brought up.

The text also explores how to apply logarithms to geography and astronomy. The section that discusses finding the distance between two places begins with the question of when it is necessary to use trigonometry. More explanations are offered than usual. Interestingly, in some of the examples logarithms are not used at all.

---

[95] Briggs, p.17.
[96] Briggs, p.29.
[97] Briggs, p.29.

The book ends with a wide variety of astronomical problems and the additional tables necessary to solve these problems. Included are various tables of the sun's declination for a variety of years and a table of the sun's right ascension in hours and minutes for every day of the year. Once again, many of these problems utilise only trigonometry. For example, one proposition asks "Having the Latitude of the place, and the Declination of the Sunne; to find the time of the Sunnes rising and setting, and the length of the day."[98] The student is instructed to add the tangent of the sun's declination to the tangent of the latitude and to subtract the radius from the sum. The remainder is then said to be the sine of the ascensional difference between the hour of six and the time of the sun's rising and setting. A numerical example is supplied, the latitude being 51 gr. 45 min. and the sun's declination 11 gr. 30 min.

9,30846,26 - the tangent of 11 gr. 30 min. - the declination
10,10328,84 - the tangent of 51 gr. 45 min. - the latitude
9,41175,10 - the sine of 14 gr. 57 min. the ascensional difference.

It was then stated that this difference being found, it "being resolved into houres and minutes, doth give a houre almost for the difference betweene the Suns rising or setting and the houre of Six, according to the time of the yeare."[99] No hint is given as to the theory behind this procedure. The astronomy problems follow the same recipe-like pattern, suggesting that the student was supposed to find the example that matched their problem closely and then to follow the steps outlined.

In general the book is very practically oriented in that no theory or explanations of why the steps suggested might work is offered at any point. The logarithms themselves are taken for granted; there is no discussion of how one would construct the tables or why logarithms work the way they do. Although a wide variety of problems are supplied as examples, there is also no exploration of why the problems can be solved in the particular ways suggested. The orientation of the book might simply be due to a perceived demand. Many people might wish to more easily solve these types of problems without displaying the slightest interest in why the procedures work. Indeed, Whiteside has suggested that even later mathematicians who read Briggs' original book were not really interested in the details of the computations.

Edmond Wingate, in his *Arithmetique Made Easie,* published in 1630, assured his readers that he had gone further than Briggs in simplifying the use of logarithms.[100] He claimed that although arithmetic was useful, many were afraid to learn it because the books that taught it were "tedious and obscure." Moreover, multiplication and division "perplex the new Practitioner, that hee takes them to be Hercules Piller, and writes upon them *Non plus ultra.*" Wingate therefore developed an instrument that used the principle of logarithms to do all manner of practical calculations.[101] His work is

---

[98] Briggs, p.44.
[99] Briggs, p.45.
[100] Edmund Wingate, *Arithmetique Made Easie*, (London, 1630), Sig. A2
[101] Wingate, Sig. A1-A4

interesting because it shows a market for vernacular texts on practical calculations based on logarithms and it adds insight to how more than one view of number could exist in a single text. He begins his explanation of numbers with Aristotelian or Euclidean definitions such as "Number is that, by which everything is numbered," with "one" described as a unit instead of a number, and numbers being multiples of units. He goes on, however, to introduce labels like "broken numbers" and "compound broken numbers" and to explain them in terms of money, weight and measure.

When Wingate turns to logarithms, he carefully distinguishes their operations from "Naturall Arithmetique", where he says that the operations are performed by the numbers themselves. Logarithms set "aside the numbers *themselves*, performe the same operations an easier way by borrowed numbers."[102] He thus separates logarithmic numbers from natural numbers; they are different in kind.[103] Many types of "numbers" existed in such practical texts side by side, although "true numbers" were still usually defined as multiples of units. The actual construction of the new "borrowed numbers" was not given a detailed explication in the text.

Only one of Napier's texts was immediately translated into English. Although Napier's *Descriptio* was immediately translated, the work being supported by an English merchant company concerned with navigation, his more theoretical *Constructio* was not translated until the 19th century. As with Briggs' work, there was no demand for the theoretical sections by the practically inclined portion of the audience who could only read English. Other English works such as John Speidell's *New Logarithms* published in 1620 also only included the tables and their uses. This work alone went through *10* editions between 1620 and 1628.[104]

CONCLUSION

One of the reasons there was no single approach to the construction of the logarithms is that until 1665, if not later, "one is hard pressed to find even a single, unified, discipline of mathematics."[105] In the late sixteenth and early seventeenth century there were many differences of opinion concerning what mathematics was or should be, the ends to which it should be pursued, the problems to be investigated, the methods to be used in solving them, and the standards of achievement.[106] Michael Mahoney has described six broad categories of mathematical practitioners who did not agree about the answers to these questions: classical geometers, cossic algebraists, applied mathematicians, mystics, artists and artisans, and the analysts. The group of practitioners that Napier is the most closely associated with are the mathematicians

---

102 Wingate, vol. 2, p. 152.

103 Wingate, vol. 2, p. 152 . His complete definition is "Logarithmes are borrowed numbers, which differ amongst themselves by Arithmeticall proportion, as the numbers that borrow them differ by Geometricall proportion." He does not give many details concerning the construction of logarithmic tables and he ignores Napier's kinematic techniques.

104 Hawkins, p.169.

105 Mahoney, *The Mathematical Career of Pierre de Fermat (1601-1665)*, p.2.

106 Mahoney, *The Mathematical Career*, p.44.

interested in practical applications such as navigational techniques including maps, charts and other applications such as fortifications and artillery. Moreover, as I have shown, Briggs also had a great concern with practical advances to navigation. As we have seen, Napier's interest in applications went beyond his mathematical research. He had wide interests in many practical activities.

Since the purpose and goal of mathematical research for practitioners such as Napier and Briggs lay in its practical applications, the applied mathematicians were more than willing to violate the canons of Greek mathematical philosophy and use techniques that we would now consider to be a part of physics. Their practical spirit overrode the classical Greek distinction between the mathematical and the mechanical. As we have seen, Napier conceived of logarithms in terms of the motions of points along lines. Euclid was ignored in that the ratios of those motions were computed in numbers and rounded off in the sixth decimal place. Napier ignored the classical distinction between continuous magnitude and discrete number. His goal was not theoretical rigor, but to create a sufficiently accurate technique for manipulating large numbers and trigonometrical functions.

The influence of logarithms on English and Scottish ideas of mathematical continua was indirect. Logarithms were so useful that they accustomed a large variety of people to working with decimal expansions of irrational numbers. The distinction made by Euclid, the separate theories of proportion for magnitudes and numbers, was not discussed. There was no available foundation for real numbers, but the utility of the logarithms necessitated that this lack of proper foundations be pushed aside. In some ways the impact of logarithms was similar to the influence of algebra. The use of algebra introduced problems whose solutions were irrational. As algebra became viewed as being extremely useful both mathematically and practically, it influenced its users to accept a wider variety of numerical solutions. The early algebra of Recorde exhibits more concern with the status and labelling of numbers that were not positive integers. But by the time of Oughtred and Harriot's work, it was merely accepted that sometimes an answer would be irrational. The status and labels of numbers became of less interest.

As we shall see when exploring the work of Wallis and Barrow, however, some mathematicians did concern themselves with the nature and status of numbers. In Napier's and Briggs' work on logarithms these issues were not directly explored so it is difficult to discern their opinions on these matters. Later scholars who addressed these issues, such as Barrow and Wallis, show that there was no real agreement or consistent treatment of numbers in this period.

CHAPTER 6

ISAAC BARROW

With Isaac Barrow, the situation changes somewhat from our earlier examples of English algebras and logarithms since his mathematical practice was different from the other scholars we have discussed. Although he was familiar with the algebraic tradition, his work was almost entirely geometrical. Moreover, he utilised symbolism when it seemed expedient, but the new symbolism was not a significant influence on his mathematical practice. Of the scholars whose work has been examined, Barrow's mathematical methods most resemble Napier's, as Barrow was also fond of using kinematic techniques. However he never occupied himself with practical problems or worked with numerical methods as Napier did. Indeed, it seems that Barrow was not particularly influenced by an interest in practical concerns. In the "Prefatory Oration" of his lectures on the usefulness of mathematical learning, however, he did recite the practical benefits of mathematics in some detail in order to encourage his students, stating,

> we may be said to receive from the *Mathematics*, the principle Delights of Life, Securities of Health, Increase of Fortune, and Conveniences of Labour: That we . . .build decent Houses for ourselves. . . .That we are protected by those Rampires from the Incursions of the Enemy; rightly use arms, skilfully range an Army. . . .That we have safe Traffick through the. . . .Sea. . . . That we rightly cast our Accounts, do Business expeditiously, dispose, tabulate, and calculate scattered Ranks of Numbers. . . .That we make pacifick Seperations of the Bounds of Lands, examine the Moments of Weights in an equal Balance. . . .[1]

But this statement has nothing to do with Barrow's actual practice; it should be viewed instead as a typical example of extolling the usefulness of mathematics so as to encourage its study.[2]

The types of problems that Barrow did solve were usually related either to geometrical characteristics of curved lines or classical geometry problems. His interest in supplying a sound foundation for numbers did not stem from a use of numerical methods, rather he wanted to subsume numbers and arithmetic under the framework of his beloved geometry.

Stevin's notion of geometric numbers also greatly influenced Barrow. He stated that since "surd" [irrational] numbers "cannot even in thought be abstracted from all

---

[1] Barrow, *The Usefulness of Mathematical Learning Explained and Demonstrated: Being Mathematical Lectures Read in the Publick Schools at the University of Cambridge*, trans. Robert Kirby (London, 1735), p. xxix.

2 Katherine Hill, "The Rhetoric of Utility: Avoiding Occult Associations for Mathematics through Profitability and Pleasure", *History of Science* **37** (1999):151-178.

Magnitude," Stevin's labelling them "Geometrical" was an excellent idea.[3] Even though Barrow seldom worked with numbers outside of a geometric context, he attempted to give irrational and rational numbers a firm foundation by associating them with magnitudes. He was also one of the only English or Scottish mathematicians in this period to discuss the nature of time and space and how these concepts were related to mathematics.

Barrow also differs from the scholars in the previous examples in that we know somewhat more about his mathematical training and education. This knowledge allows us some explanation of the direction of his research. Secondly, Barrow actually discussed his views about the nature of numbers and mathematical continua in some detail in a series of lectures given at Cambridge. Although Barrow is often depicted as a traditional figure because of his rigorous geometric style, his opinions on these subjects were very different from the classical conceptions. Indeed, Barrow supplies us with our first extended discussion of the nature of numbers and continua, and his work shows us how far the English and Scottish mathematicians were able to move from Greek ideas.

This chapter will first briefly investigate Barrow's education and mathematical training. We will then have some idea about what influenced his particular opinions beyond the general ideas we discussed in the first chapter. Secondly, we will explore portions of Barrow's actual mathematical practice through a discussion of some of the mathematical problems presented in his translation of Euclid's *Elements* and in his *Geometrical Lectures*. This will supply us with a picture of Barrow's mathematical research which will also help to explain why he took the positions he did regarding the nature of numbers and their relation to abstract magnitude. Finally, we will examine his Cambridge lectures on the foundations of mathematics to show his specific ideas relating to these topics. As we will see in the next chapter on John Wallis, however, Barrow's notions were not universally accepted.

BACKGROUND AND EDUCATION

The education given to those bound for university in English secondary schools concentrated on Latin grammar. In 1677, John Newton remarked that he had never heard of any grammar school in England in which mathematics was taught. The new demands on surveying and navigation ensured that young men learning a trade obtained an increasingly complex mathematical education.[4] University instruction in mathematics became the focus of increased attention due to outside pressure for reform coupled with inside pressure from movements such as Ramism.[5] An investigation of the

---

[3] Barrow, *The Usefulness of*, p.45.

[4] E. G. R. Taylor, *The Mathematical Practitioners of Tudor and Stuart England* (Cambridge: Cambridge University Press, 1970), p. 3. See also Katherine Hill, "Mathematics as a Tool for Social Change: Educational Reform in Seventeenth-century England", *The Seventeenth Century* **12** (1997):23-36.

[5] Ramism was associated with Puritanism and regarded by conservatives as a dangerous development. Ramus attacked almost every aspect of Aristotelian thought, including the *Physics*; he was also critical of Euclid. He wanted to reform the curriculum and teaching methods. See Walter J. Ong, *Ramus: Method and the Decay of*

mathematical education given Barrow will provide a glimpse of a system in transition.

Figure 1.

Barrow's family were royalists: indeed, his father joined the exiled court in France during the Interregnum. After an unsuccessful attempt to board Barrow at another school, his family decided to send Barrow, to Felsted School in Essex, a Puritan and Parliamentary stronghold. The master of Felsted, Martin Holbeach, with whom both Wallis and Barrow studied, was noted for his Puritan sympathies: it was said of him that

---

Dialogue (Cambridge Mass.: Harvard University Press, 1983), pp. 32-33. Wilber Samuel Howell, *Logic and Rhetoric in England, 1500-1700* (New York: Russell and Russell, 1956), pp. 146-247. John Gascoigne, "A Reappraisal of the Role of the Universities in the Scientific Revolution" in *Reappraisal's of the Scientific Revolution* ed. David C. Lindberg, Robert S. Westman (Cambridge: Cambridge University Press, 1990), pp207-260, see pages 223-24.

he "scarce bred any man that was loyall to his prince."[6] Nonetheless, the school was more famous for its stern discipline and classical curriculum than for its political leanings. We know from Wallis' autobiography that they studied at Felsted not only Latin and Greek, but Hebrew and a smattering of Logic as well.[7] One can assume that Barrow's education was similar and that he learned no mathematics at all at Felsted.

In December of 1643, Barrow was to leave Felsted for the Royalist Peterhouse College, Cambridge. His uncle, who was a Fellow at this college, was expected to take charge of his education. But this was a time of political upheaval in Cambridge as Cromwell occupied it in 1643 and half of the university Fellows were purged and replaced by Parliament. Parliamentary visitors expelled his uncle before Barrow was able to arrive at Cambridge. It was not until 1646 that he entered Trinity College, Cambridge.[8] Although Trinity was a very Presbyterian college, Barrow survived by receiving financial support from the Walpole family and intellectual support from his tutor, James Dupont. Dupont was a Royalist Anglican who had strong scholastic leanings.[9] No direct information exists on the course of Barrow's studies, except that he continued his education in Latin, Greek, and Hebrew.

Barrow had, however, opportunities to study Descartes' work. John Smith of Queen's College gave lectures on mathematics, specifically on Descartes's *Geometry*, in 1648 or 1649.[10] Moreover, Descartes' work was embraced for a time by the Cambridge Platonists Henry More and Ralph Cudworth. However, historians may have exaggerated the amount of unanimity amongst the Cambridge Platonists, and it is not clear what impact they had on the curriculum. Moreover, the evidence that could document the course of Barrow's particular mathematical education is fragmentary at best, but most students in their second or third year took a course consisting of arithmetic, geometry and optics.[11] We know, however, that in 1652 Barrow chose as the subject of his M.A. thesis "That the Cartesian Hypothesis Concerning Matter and Motion Scarcely Satisfies the Phenomena of Nature."[12] Although Barrow praised Descartes' "extraordinary skill as a mathematician" and conceded that his philosophy offered a reasonable explanation of natural phenomena, he nevertheless had grave

---

[6] Mordechai Feingold, "Isaac Barrow: Divine, Scholar, Mathematician," in *Before Newton: The Life and Times of Isaac Barrow* (Cambridge: Cambridge University Press, 1990), p. 2.
[7] Christoph Scriba, ed. "The Autobiography of John Wallis, F.R.S.," *Notes and Records. Royal Society of London* 25 (1970): 27. Wallis and Barrow both attended Felsted, although not at the same time.
[8] Feingold, "Isaac Barrow," p.4.
[9] John Gascoigne, "Isaac Barrow's Academic Milieu," in ed. Mordechai Feingold, *Before Newton: The Life and Times of Isaac Barrow* (Cambridge: Cambridge University Press, 1990), pp. 250-290. Also, Scholasticism underwent a revival at Cambridge, perhaps because it helped Conservatives fight off attacks based on the word of God. See Kearney, pp. 81-84.
[10] Feingold, "Isaac Barrow," pp. 19-21.
[11] *The Works of the Learned Isaac Barrow, D.D. Late Master of Trinity College in Cambridge (Being all his English Works)*, published by his Grace Dr. John Tillotson, the late Archbishop of Canterbury (London, 1716) contains an account of Barrow's life by A. Hill which states, " When he was yet a young scholar . . . he applied himself to the reading and considering the Writings of Lord Verulam, M. des Cartes, Galileo, and the other great Wits of the last age." He also claims that Barrow studied classical mathematics with Mr. John Ray.
[12] Gascoigne, p.264.

reservations about explaining nature in purely mechanical terms. By 1652 Barrow had not only read Descartes' mathematical work, but also enough other mathematics to make him appreciative of Descartes's mathematical techniques. He did, however, offer Aristotle and Plato as alternatives to Cartesian mechanism.[13] Interestingly, Barrow used Bacon's ideas about the importance of empirical data to criticise Descartes' dependence upon metaphysical truths. By 1655, his patron James Worthington wrote of him,

> [Barrow] had no body to instruct him in Mathematics, but *proprio Marte*, he conquered all difficulties in the most crabbed authors. . . . He looks upon his performance upon Euclid as a small business to him. He knows Tacquet &c...[14]

Apparently, Barrow was able to teach himself both classical and contemporary mathematics.

Barrow's academic career was interrupted during the Interregnum. In 1653, a year after Barrow received his M.A., Parliament "considered among themselves of the suppressing Universities and all Schools of Learning, as heathenish and unnecessary."[15] Even after the immediate threat of closing the universities was over, their curriculum was under attack: in 1654, John Webster's *The Examination of the Academies* asked, "Can the Mathematical Sciences, the most noble, useful, and of the greatest certitude of all the rest, serve for no more profitable end, than speculative processes?"[16] As a Royalist, Barrow was in a very precarious position, especially after his tutor, James Dupont, was purged in 1654. A scheme was hatched for Barrow to replace Dupont as the Regius Professor of Greek, but it failed. During this period Barrow began a serious study of mathematics and composed a series of treatises on certain portions of Greek mathematics. For example, he worked on his first publication, *Euclidis elementorum libri xv breviter demonstrati*, as well as his edition of Archimedes.[17] The insecurity of Barrow's position in this period led him to travel on the Continent in 1655. First travelling to France and Florence, he was attacked by pirates on his way to Constantinople. Unluckily, the ship taking him to Venice on his way home to England caught fire, and Barrow lost all his possessions. [18]

By September 1659, he was back in Cambridge and Bishop Ralph Browrigg soon ordained him. The Restoration occurred in the next year and Barrow was rewarded with the Regius professorship of Greek that he was denied before.[19] In 1662, he also became the Gresham Professor of Geometry. The new duties were light, as he had to

---

[13] Percey H. Osmond, *Isaac Barrow: His Life and Times* (London: the Society for Promoting Christian Knowledge, 1944). He quotes Barrow's M.A. thesis as saying "I confront him [Descartes] with Plato, Aristotle, Democritus or Epicurus..." page 30. Barrow seems at this point to have had a deep respect for classical sources, and a reluctance to dismiss scholastic learning.

[14] Feingold, "Isaac Barrow," p. 39.

[15] Richard Foster Jones, *Ancients and Moderns: A Study of the Background of the Battle of the Books* (St. Louis: Washington University Press, 1936), p. 119.

[16] Jones, p. 259.

[17] Besides his work with Euclid's *Elements* and Archimedes' *De sphaera & cylindro* Barrow also composed manuscripts on Apollonius' *Conics* and the *Sphaerics* of Theodosius, all by the end of November 1653. See Feingold, pp.44-45.

18 Feingold, "Issac Barrow", pp.47-53.

[19] Whiteside, "Barrow, Isaac," *DSB*, p.473.

lecture on arithmetic and geometry one day a week. During this period, Barrow became a member of the Royal Society, but he never contributed reports or attended meetings. He was able to give up the Regius and Gresham professorships and reduce his teaching load when he became the first Lucasian professor of mathematics at Cambridge in 1663.[20]

Anthony Grafton, in his article "Barrow as a Scholar," argues that Barrow should be seen as belonging to four distinct intellectual traditions that studied the classics in early modern England. First, there was a pedagogical tradition that concentrated on helping students though classical texts. Barrow's translations of Euclid and Archimedes can be seen as part of this effort. Second, there was an Aristotelian tradition, in which English scholars such as John Case based university education in all fields as far as possible on Aristotle's texts and methods. Barrow's respect for Aristotle coupled with the recommendation of Aristotle by his tutor, James Dupont, to his students show that Barrow was educated within this Aristotelian tradition. Third, there was an encyclopedic tradition that desired to fuse the Greek and the Biblical traditions. Barrow was drawn to this study by his love of classical philosophy and literature together with his love of theology, although he made no original contributions. Fourth, there was a tradition that studied the history of the sciences in antiquity. Henry Savile, for example, had undertaken a systematic effort to collect, study and explain the main ancient texts in the exact sciences.[21]

Barrow's links with the first two traditions were official. He not only had a post teaching the humanities, but he himself wrote "a deliberately artistic and elaborate Latin prose and verse on many ceremonial occasions." Moreover, Barrow also argued for the value of the ancients as the basis of a coherent education, and in particular for Aristotle's unique "pedagogical and intellectual value as the basis for university teaching and disputation."[22] In many ways Barrow belonged both to the humanistic and Aristotelian traditions.

Barrow was not, however, only familiar with the humanist and classical traditions. At the beginning of his professional mathematical career, Barrow had taught himself not only classical mathematics, but also the modern work of Descartes.[23] He had translated not only Euclid's *Elements* but also portions of Archimedes, Apollonius and Theodosius, although this later part of his work was not published until 1675. Besides Descartes, he had also read Torricelli's *Opera geometrica*, which included a method of tangents.[24] Archimedes seems to have been an important influence on Barrow, as he

---

[20] Whiteside, *DSB*, p.473.

[21] Anthony Grafton, "Barrow as a Scholar," *Before Newton: The Life and Times of Isaac Barrow*, ed. Mordechai Feingold (Cambridge: Cambridge University Press, 1990), pp.291-295.

[22] Grafton, p. 295.

[23] Mahoney suggests in "Barrow's Mathematics: Between Ancients and Moderns," *Before Newton: The Life and Times of Isaac Barrow*, ed. Mordechai Feingold (Cambridge: Cambridge University Press, 1990), p. 179, that at the beginning of Barrow's professional mathematical career he was familiar with the corpus of ancient mathematics as well as the writings of Mersenne, Descartes, Pascal, Huygens, Viviani, Torricelli, Renaldini, Snel, Alsted and Kersey.

[24] Feingold, "Isaac Barrow," pp. 45-46. The manuscript consisted of the contraction of most of the know works of Archimedes, the first four books of Apollonius' *Conics*, and the *Sphaerics* of Theodosius.

followed Archimedes (and perhaps Descartes) by constructing curves by using the idea of a point moving though space. But in Barrow's edition of Archimedes he rewrote the ratio theory proofs forms of the original Greek into "free variable notation" and he doubted when he came to particularly involved forms that Archimedes was describing his actual method of discovery. He supposed, instead, that Archimedes had a method of analysis much closer to the "modern" form.[25] Although Barrow preferred traditional geometric techniques he was not adverse to introducing symbols to make comprehension easier by simplifying the visual layout as he did with his versions of Euclid and Archimedes.

Barrow was also familiar with the second book of Descartes' *Geometrie* where Descartes constructed curves by using points moving through space. He did not, however, emphasise special rulers or instruments that would be useful in the construction of curves as Descartes had. In later sections of this work dealing with tangents and areas, Barrow chose points of reference that best suited each individual problem, nor did he ever utilise Descartes' notation. Yet in Lecture IV of his *Geometrical Lectures,* he did treat the properties of curves generated by the composition of a uniform horizontal motion and a constantly increasing downward motion. In this lecture he discussed a Cartesian theme; the generality of these methods. Barrow also cited Descartes' mathematics approvingly in his Cambridge mathematical lectures. It should, however, be kept in mind that Barrow's attitude towards Descartes had changed from qualified respect to restrained hostility by 1661, due to the growing influence of Hobbes' work and Barrow's belief that Descartes' work had the same materialistic tendencies as Hobbes'.[26] In general, although Barrow was familiar with his contemporaries' mathematical practice, he did not always utilise their methods.

Barrow was educated in the classical tradition. Through his tutor, Dupont, he had a strong knowledge of the Greek language and of ancient philosophy. In general, he also preferred the classical geometric methods. He believed that the importance of accepting authority, in both religion and science, had been shown by the "anarchy" of the Civil War.[27] Even when disagreeing with Aristotle, he said "I speak with all due Regard to the Honour of that great Man, for whom, notwithstanding, I preserve the greatest Veneration and Esteem, by reason of his most profound Wisdom and extensive Learning. . ."[28] Although he often disregarded classical concepts, in many ways he attempted to remain in what *he* considered to be the traditional boundaries of mathematics.

BARROW'S MATHEMATICAL PRACTICE

Barrow felt so strongly about the merits of geometry that he explicitly denied that arithmetic had an existence independent of geometry, stating:

Unfortunately, Barrow's mathematical notebooks have not survived.

[25] Whiteside, "Patterns of Mathematical Thought," pp. 191-192.

[26] Gascoigne, pp. 277-278.

[27] Gascoigne, "Isaac Barrow's Academic Milieu," pp. 261-262.

[28] Barrow, *The Usefulness of Mathematical Learning*, p. 119.

> I affirm the Whole of mathematics to be in some sort contained and circumscribed within the Bounds of Geometry. . . . For what science can be more accurate than geometry? What Science can afford Principles more evident, more certain, yea I will add, more simple, than Geometrical Axioms?[29]

Arithmetic, then, was subservient to geometry in Barrow's scheme of mathematics. Moreover, Barrow claimed to be merely *restoring* arithmetic to its proper place, although he could, of course, cite no examples of a classical source that agreed with his point of view.[30] One reason he subsumed arithmetic into geometry is that he believed he was vindicating Euclid "the *Father* and *Prince* of Geometry from the undeserved Reproaches which are every where thrown upon him." Ramists, for example, and some mathematicians like John Wallis, thought Books II and V of the *Elements* were merely poorly done arithmetic. Barrow's hostility towards this idea caused him frequently to attack Ramus in his mathematical lectures, and he also criticised Wallis, although more politely.[31]

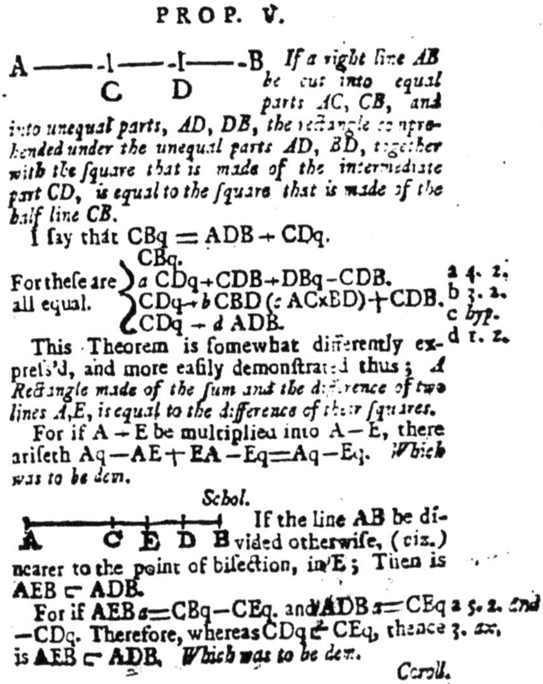
PROP. V.

A——C——D——B *If a right line AB be cut into equal parts AC, CB, and into unequal parts, AD, DB, the rectangle comprehended under the unequal parts AD, BD, together with the square that is made of the intermediate part CD, is equal to the square that is made of the half line CB.*

I say that CBq = ADB + CDq.

For these are all equal.
CBq.
a CDq + CDB + DBq − CDB.
CDq + b CBD (c ACxBD) + CDB.
CDq + d ADB.

a 4. 2.
b 3. 2.
c hyp.
d 1. 2.

This Theorem is somewhat differently express'd, and more easily demonstrated thus; *A Rectangle made of the sum and the difference of two lines A, E, is equal to the difference of their squares.*

For if A + E be multiplied into A − E, there ariseth Aq − AE + EA − Eq = Aq − Eq. *Which was to be dem.*

Schol.

A C E D B If the line AB be divided otherwise, (viz.) nearer to the point of bisection, in E; Then is AEB ⊏ ADB.

For if AEB a = CBq − CEq. and ADB = CEq − CDq. Therefore, whereas CDq ⊏ CEq, thence is AEB ⊏ ADB. *Which was to be dem.*

a 5. 2. *and* 3. *ax.*

Coroll.

[32]

Figure 1

[29] Barrow, *The Usefulness of Mathematical Learning*, p. 46.
[30] Barrow, *The Usefulness of Mathematical Learning*, p. 29.
[31] Barrow, *The Usefulness of Mathematical Learning*, pp. 37, 54, 124, 331. There might have been a religious and political component to this disagreement; as previously mentioned Raminism was associated with Puritanism and reform. Barrow was strongly opposed to both.
[32] Barrow, *Euclide's Elements*, p. 43.

Barrow's translation of Euclid's *Elements* (especially Book II) provides us with the viewpoint of a seventeenth-century mathematician who favoured geometrical methods. Since Book II was often considered to be geometrical algebra in this period, Barrow's treatment of its propositions provides hints as to his opinion of the relationship between unknowns, numbers, and geometrical magnitudes at an early stage of his career. Barrow claimed to have two ends to accomplish in his translation of the *Elements*: to be perspicuous and brief. Euclid's own demonstrations, he claimed, were used in condensed form. The second book is stated not to ". . . deviate in any particular from him."[33] In order to "content the Desires of those who are delighted more with symbolical than verbal Demonstrations,"[34] he used the symbols of Oughtred. The first proposition we will explore is Barrow's version of Book II proposition 5, which is different than the version of the proof given in Heath's translation of Euclid. He began by drawing a line that is divided into four parts.[35] He stated that $BCq = ADB + CDq$, (with ADB standing for the rectangle made from AD and DB) and proves this in the following fashion. These equations are said to be equal:

$$CBq \tag{1}$$

$$CDq + CDB + DBq + CDB \tag{2}$$

$$CDq + CBD\ (AC \cdot BD) + CDB \tag{3}$$

$$CDq + ADB^{36} \tag{4}$$

Equation (2) was justified by Book II proposition *4* which he represented as $Zq = Aq + Eq + 2AE$. Equation (3) was justified by Book II proposition *3* which stated that $ZE = AE + Eq$, and also by hypothesis, while equation (4) was justified by an appeal to Book II proposition *1*. This particular proof appears to be algebraic in so far as it is justified and given in terms of equations. Also, operative symbolism is in use. Mathematical relations and not objects seem to be the key factor. He makes no appeal to an "intuitive ontology of the physical world."[37] All these factors would seem to point to the algebraic nature of his proof. Barrow himself, however, thought he had strictly followed Euclid's own proofs in his rendition of Book II. He seemed, however, to have no qualms about associating unknowns and magnitudes.

Barrow tended to move back and forth between "algebraic" and "geometric" forms

---

[33] Isaac Barrow, *Euclide's Elements; The Whole Fifteen Books Compendiously Demonstrated with Archimedes Theorems of Sphere and Cylinder, investigated by the Method of Indivisibles* (London: 1714), preface.

[34] Barrow, *Euclide's Elements*, preface.

[35] Barrow, *Euclide's Elements*, p. 43.

[36] Barrow, *Euclide's Elements*, p. 43.

[37] Mahoney, "The Beginnings of Algebraic Thought in the Seventeenth Century," p. 142. He defines the algebraic mode of thought as having three main characteristics. Mahoney states, "First, this mode of thought is characterised by the use of an operative symbolism. . . .Second,..the algebraic mode of thought deals with mathematical relations rather than objects. . . . .Third, the algebraic mode of thought is free of ontological commitment."

of proof, between using equations and numerical examples and geometrical construction. His proof of proposition 1 Book II used construction and justification in the manner of Euclid. In a Scholium to II 1, however, he lapsed back into an explanation in terms of equations and spoke of products in the sense of modern multiplication. He cautioned the reader to take particular care of signs while multiplying lines.[38] He discussed

> . . . the nine following propositions . . . arising from comparing of lines multiplied into themselves (which you may find done to your hand in Vieta and other Analytical Writers) are demonstrated with great facility, by reducing the matter for the part to almost a simple work.[39]

At this stage of his career he appears to have believed that multiplication in the manner of Viète and the "analytical writers" could be seen as a device for the simplification of proofs.[40] Analytic was usually used by Barrow to designate approaches we would refer to as algebraic. At times, however, Barrow seems to have used analytic to refer to methods of indivisibles such as Cavalieri's. Some of his proofs were given both by equations and by construction. II *4* is first proved in an "algebraic" fashion and then by construction. It is difficult to ascertain why in some cases Barrow felt a need to give two types of demonstrations. Indeed, at some points, such as II *12* in the scholium, Barrow used specific numbers, perhaps illustrating how the equations might be solved.[41] Barrow used numerical examples in some problems though not in others. Overall, Barrow's version of Book II varied in its approach to proofs. No overarching scheme seemed to exist. The most puzzling aspect is his claim that he followed Euclid's work in Book II precisely. If he had seen any difference between the "algebraic" formulation and the "geometric", he would not have made this claim.

Although Barrow was willing to use an "analytic" formulation in his translation of Book II, meaning he at least utilised equations, there is no indication that he believed he was dealing with anything other than geometry. All mathematics was geometry for Barrow. His desire for brevity never decreased his demand for rigour. It appears that his use of symbolism and numeric examples in his version of Euclid was mainly intended as an aid to understanding for his readers. But like Descartes and Viète, he was willing to interchange numbers, unknowns, and geometric magnitudes.

Before moving on to discuss Barrow's *Geometrical Lectures* we should briefly examine the allegation that everything important in this work was borrowed from James Gregory's *Geometria pars universalis*, published in 1668, two years prior to Barrow's book. Joseph Hofmann and D.T. Whiteside have defended this point of view.[42] Mordechai Feingold has responded with an article that traces the publishing history of

---

[38] Barrow, *Euclide's Elements,* p. 40.

[39] Barrow, *Euclide's Elements*, p. 41.

[40] For example, Heath's version of II *2*, although not excessively long (half a page) is reduced by Barrow to three lines of equations. The same justification is used in both proofs. Barrow may have perceived no difference between his proof and Euclid's.

[41] Barrow, *Euclide's Elements*, p. 47.

[42] Joseph Hofmann, *Leibniz in Paris, 1672-1676: His Growth to Mathematical Maturity* (Cambridge: Cambridge University Press, 1974), p. 104 ; Whiteside, "Patterns of Mathematical Thought," p.368, n. 9.

Barrow's book.[43] Feingold agrees that Barrow owned a copy of Gregory's *Geometria* before 1670, but he argues that by the time Barrow received Gregory's book, his own work was almost finished and in the hands of John Collins.[44] Futhermore, he claims that what borrowings do exist are related to Collins's request that Barrow elaborate on a few of Gregory's theorems. Feingold also points out that Gregory never suggested that Barrow was indebted to him; Gregory seems to have respected and admired Barrow's mathematical abilities. It seems that the similarities that do exist between Barrow's geometrical lectures and Gregory's *Geometria* are due to the fact that they drew upon similar traditions and worked upon related topics.

Barrow's *Geometrical Lectures* have been discussed in detail in Michael Mahoney's article "Barrow's Mathematics: Between Ancients and Moderns," but a brief overview of his lectures and a few examples of the types of problems he solved will supply us with an idea of Barrow's mature mathematical style. Barrow's *Geometrical Lectures* deal with the geometrical characteristics of curved lines. The preface claims that he did not originally intend these lectures for the public, but that they were school lectures which were "read by Duty." It begins by embedding the properties of curves in the larger context of the generation of magnitude. The first lecture treats the generation of magnitude, which he believes to be the origin of mathematics.[45] He goes on in the second lecture to discuss the generation by motion of geometrical shapes such as cylinders, parabolas, and hyperbolas.[46] The third lecture is devoted to compound motions and the generation of such figures as the Spiral of Archimedes and the Quadratrix.

After providing a basic framework for discussion, Barrow explored the kinematic properties of curves. In Lecture IV he discussed the properties of curves generated by the composition of a uniform horizontal motion and a constantly increasing downward motion. In this section Barrow did not utilise infinitesimal or limit techniques. He went on in Lecture V with his analysis of the properties of curves generated by the composition of uniform and accelerated motion, concentrating especially on tangents, normals, and extreme values.[47] Lectures I - V were a survey of the most recent developments in mathematics. In Lecture VI, however, Barrow moved onto material that was less well known. He began an "Investigation and ready Demonstration of Tangents, without the trouble of Calculation . . . [which] have not been so fully handled or exhausted as some others . . ."[48] This lecture deals mainly with the properties of Conics. Except for the kinematic aspects of the demonstrations, Barrow treats mainly

---

[43] Mordechai Feingold, "Newton, Leibniz, and Barrow Too: An Attempt at a Reinterpretation," *Isis* 84 (1993): 310-338.

[44] Feingold, "Newton," p.333.

[45] Barrow's opinions on time, space and the composition of lines will be discussed in the section of the chapter concerning continua.

[46] Isaac Barrow, *Geometrical Lectures: Explaining the Generation, Nature and Properties of Curve Lines*, trans. by Edmund Stone (London: Printed for Stephen Austen, at the *Angel* and *Bible* in St. Pauls Church-Yard, 1735), p.35.

[47] See Mahoney, "Barrow's Mathematics," pp. 207-213 for an extended discussion of this section.

[48] Barrow, *Geometrical Lectures* , p.100.

the properties of curves discussed in Apollonius' *Conics*.[49]

Proposition 14 is fairly illustrative of the work done in this section.

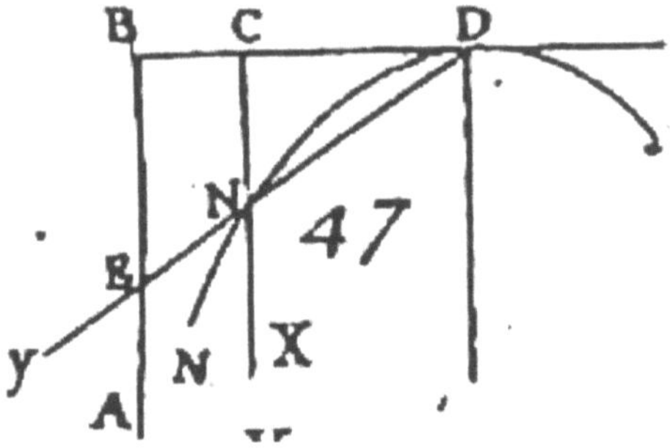

[50]

Figure 3.

He let the straight lines *DB, BA* be given in position, and let the straight line *CX* move along the straight line *DB* while remaining parallel to *BA;* at the same time he also let a straight line *DY* revolve around the point *D* so that it cut *BA* in *E*. He also defined *R* by supposing that there is always the same ratio between *BE, DC*, which he assumed was the same ratio as some quantity *R* had to *DB*. Next, he let the right lines *DE, CX* intersect one another at the point *N*. He concluded that the "line" *DNN* will be a parabola. The proof is geometric.

$$\text{For let } R : DB = DB : P. \qquad (5)$$

Barrow here introduces the quantity *P* which is defined by the above relationship, although he could have defined *P* directly.

$$BE : DC = DB : P. \qquad (6)$$

This follows from the definition of *R* given in the construction of the problem.

$$DB : BE = DC : CN. \qquad (7)$$

By the similar triangles *DBE* and *DCN*.

$$DB : BE + BE : DC = DC : CN + DB : P. \qquad (8)$$

By adding *6* and 7

---

[49] Mahoney, p. 212.

[50] Barrow, *Geometrical Lectures*, p. 181. Plate 5.

$$DB : DC = DC \cdot DB : CN \cdot P \qquad (9)$$

By multiplying 6 and 7

$$DB \cdot DC : DC\,q = DC \cdot DB : CN \cdot P. \qquad (10)$$

Multiply *9* by *DC/DC*, on the left.

$$.\ DCq = CN \cdot P. \qquad (11)$$

"Whence it is manifest that the curve *DNN* is a Parabola, whose Parameter is *P*, Vertex *D*, and Diameter parallel to *BA*.[51] The last step is based on the ordinate properties of the parabola as defined by Apollonius. Barrow stated that the earlier propositions in Lecture VI, including Proposition *14*, were "perhaps common: But what follows cannot be said to be such."[52] In the previous example he was not claiming to offer new techniques. He certainly avoided "calculations" in this proposition; he depended upon geometrical properties such as similar triangles to complete his proof. He differed from classical demonstrations, such as Apollonius', which were static, by using lines or points moving though space. Occasionally, he utilised symbols and some of his proofs seem on the surface to be as algebraic as some of proofs he gave for the propositions of Book II of Euclid's *Elements*. He was, however, careful to keep the terms of his 'equations' in second degree to assure geometric intelligibility.

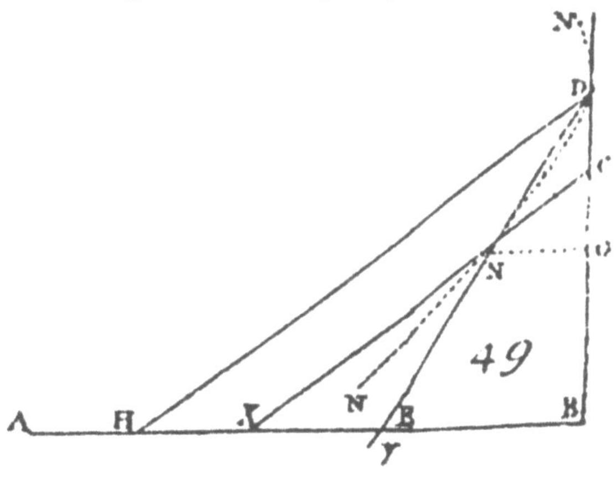

Figure 4

[51] Barrow, *Geometrical Lectures*, p.110. Although I have attempted to keep to the spirit of Barrow's proof, some adaptations have been made. For example he used *A.B::C.D* to express *A* bears to *B* the same ratio as *C* to *D*. Also the justifications given for each step were not supplied by Barrow, he merely provided the 'equations.'

[52] Barrow, *Geometrical Lectures*, p. 112.

In Proposition *16*, which superficially seems more algebraic, we still have *DB,BA* given in position as in Proposition *14*, but now we have the straight line *CX* moving parallel to *DH* which is also given in position. Once again, Barrow defined *R* by supposing that there is always the same ratio between *BE, DC*, which he assumed was the same ratio as some quantity *R* had to *DB*. Barrow states that "the Intersection *N* will be a hyperbola." He draws *NG* parallel to *BA* (in effect setting up a coordinate system) and makes some substitutions, perhaps, in order to make his demonstration easier to follow.

$DB = b;\ BH = h;\ DG = x;\ GN = y$. Then

$$. x : y = b : by/x = BE.^{53} \tag{11}$$

By the ratios of the triangle sides and similar triangles.

$$h : b = y : by/h = GC. \tag{12}$$

Again using the ratios of the triangle sides and similar triangles.

$$CD = x - by/h \tag{13}$$

and then

$$by/x : x - by/h = b : r \tag{14}$$

by supposition. At this point Barrow might have manipulated his equations in the following way, but he did not provided these steps.

$$(by/x)/(x - (by/h)) = b/r \tag{15}$$

$$(byr)/(xb) = x - by/h \tag{16}$$

$$(byr)/xb = (hx - by)/h \tag{17}$$

$$byrh = bhx^2 - b^2xy \tag{18}$$

$$yx + (yrh)/b = (h/b)x^2 \tag{19}$$

Barrow merely gives us

$$yx + hry/b = (h/b)\ xx \tag{20}$$

He then substitutes $hr/b = m$ to finish with

[53] It is not clear why he labels some lines and not others.

$$yx + my = (m/r)xx, \tag{21}$$

which he had shown to be a hyperbola in proposition *10*. Therefore the curve *DNN* is a hyperbola.[54] It seems that whatever analytical tools Barrow used were mainly for the sake of expediency. One can suppose that he manipulated the equations in an algebraic fashion to move from step *4* to step *5*, but he mainly relied upon geometrical techniques and lines moving through space to prove his proposition.

Nevertheless, we can see from Lecture X that Barrow was familiar with a wide range of techniques. He gave a method of finding tangents similar to Fermat's.[55] He realised that this type of method was "common and published" but he was advised to include this portion by a friend.[56]

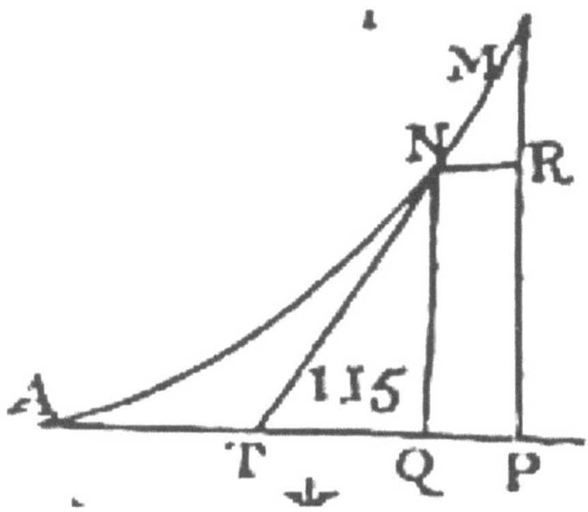

57

Figure 5.

He let *AP, PM* be straight lines given in position and let *PM* cut the proposed curve at *M;* let *MT* touch the curve at *N* and cut *AP* at the point *T.* To find the length of *PT* (the subtangent) he supposed that the arch of *MN* was "indefinitely small."[58] Next, he drew the straight lines *NQ, NR* parallel to *MP, AP.* He then let $MP = m$, $PT = t$, $MR = a$, and $NR = e$. He was now ready to "compare *MR,NR* expressed by Calculation in an Equation," observing these rules:[59]

1. In the computation, reject all terms where a power of *a* or *e* or their product appears because "these Terms will be equal to nothing."

---

[54] Barrow, *Geometric Lectures*, p.112.
[55] Mahoney, "Barrow's Mathematics," p. 266.
[56] Barrow, *Geometric Lectures*, p. 172.
[57] Barrow, *Geometric Lectures*, p. 181. Plate 5.
[58] Barrow, *Geometric Lectures*, p.173.
[59] Barrow, *Geometric Lectures*, pp. 173-174.

2. After the equation is set up, reject all terms consisting of known or determinate quantities, or terms where *a* or *e* do not occur because "these Terms will be equal to nothing."

3. Substitute *m* (or *MP*) for *a*, and *t* (or *PT*) for *e*, "by which means the Quantity of *PT* will be found.
And he concludes with,

> When any indefinitely small Particle of the Curve enters the calculation, I substitute in its stead a Particle of the Curve properly taken; or any straight line equal to it, because of the indefinitely Smallness of the Part of the Curve.[60]

Barrow did not supply a derivation for this method or suggest any theoretical foundations[61] but he did offer some examples. The second example dealt with a "cubic circle" $x^3 + y^3 = r^3$.

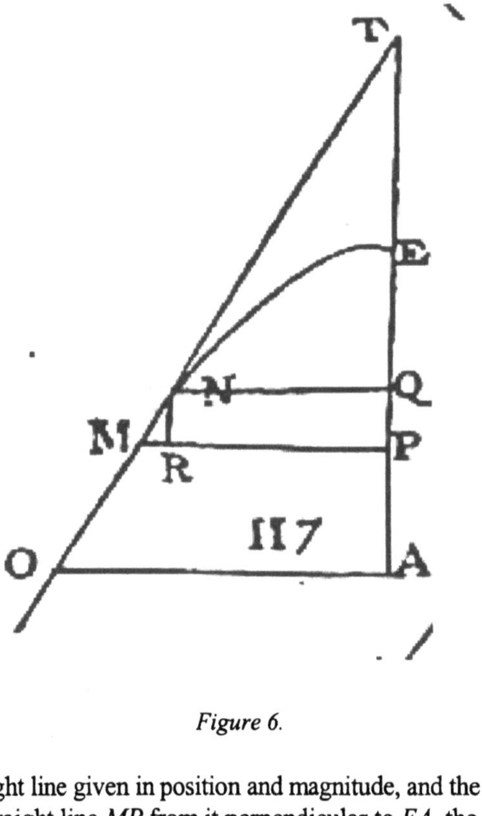

62

Figure 6.

He let *EA* be a straight line given in position and magnitude, and the curve *EMO* be such that drawing any straight line *MP* from it perpendicular to *EA*, the sum of the cubes of

[60] Barrow, *Geometric Lectures*, p.174.
[61] See Mahoney, "Barrow's Mathematics" pages 225 - 226 for a explanation of the method.
[62] Barrow, *Geometric Lectures*, p. 181. Plate 5.

AP and *MP* are equal to the cube of the line *AE*.[63] He also let $AE = e$ and $AP = f$. The proof given follows the above mentioned rules.

$$AQ = f + e \tag{22}$$

By supposition:

$$AQ^3 = f^3 + 3f^2e + 3fe^2 + e^3 \tag{23}$$

By multiplication.[64]

$$AQ^3 = f^3 + 3f^2e \tag{24}$$

By using the rule to drop "superfluities."

$$NQ = m - a \tag{25}$$

By supposition, since MP = m and MR = a.

$$NQ^3 = m^3 - 3m^2a + 3ma^2 - a^3 \tag{26}$$

By multiplication.

$$NQ^3 = m^3 - 3m^2a \tag{27}$$

Again by the rule to drop "superfluities".

$$AQ^3 + NQ^3 = AE^3 = f^3 + 3f^2e + m^3 - 3m^2a = r^3 \tag{28}$$

By adding, and by supposition.

$$3f^2e = 3m^2a \tag{29}$$

By the rules again, putting m and t for a and e.

$$.f^2t = m^3 \tag{30}$$

Replacing a and e *with* m and t as part 3 of the rule requires.

$$t = m^3/f^2. \tag{31}$$

---

[63] Barrow, *Geometric Lectures*, p.175.

[64] Barrow wrote, *AE* cub. $= f^3 + 3ffe + 3fee + e^3$. He occasionally used "cub." instead of exponential notation.

"therefore *PT* is a fourth Proportional in the continued ratio of *AP* to *PM*."[65]
He then generalised by noting that if you let $AP^2 + MP^2 = AE^2$, you will find *PT* to be m^4 / f^3 "and so on."[66] This problem shows that Barrow was familiar with Fermat's techniques. Nevertheless, he preferred to utilise geometry.

These examples of Barrow's mathematical practice show that he did not often directly utilise numerical techniques, arithmetic, or algebra, although he was familiar with the latest methods. He favoured geometry and valued rigour in mathematics. None of his work depended upon numerical properties and he seldom overtly used algebraic manipulations. Nonetheless, in his Cambridge lectures Barrow supplied a new foundation for numbers. As we will see, he seems to have been mainly motivated by his desire to show that all mathematics, and indeed all of science, should be subsumed under geometry.

FOUNDATIONAL ISSUES IN BARROW'S CAMBRIDGE LECTURES

Before turning to the content of Barrow's lectures, it is important to note that they were directed towards an undergraduate audience who might have had some knowledge of Euclid's *Elements* and the fundamentals of arithmetic. Barrow's aim was not to teach his student's how to do mathematics, but even so, it was apparently slow going. Almost midway through the second year of the lectures, Barrow lamented:

> But shall I never extricate myself from these Quirks and Trifles? Shall I always spend my time in examining what is of no Value? In so plentiful a Harvest, so rich a Vintage, so great a Store of most important Disquisitions, why do I only glean the scattered Ears, search the neglected Boughs, and gather the fallen Grapes? When a Chase after the important and difficult Things in the Mathematics is offered, a Chase so full of Variety, so pleasant and so certain; wherefore do I dwell so long upon these little Questions, like one hunting after Flies?[67]

Barrow went on in this vein for some while, showing what must have been a genuine impatience. It was not only the slow pace that was the problem, Barrow could never be sure of his audiences' understanding. Some of the below distinctions that Barrow makes, in particular in regard to his remarks on the nature of space seem quite confusing. This confusion, however, might stem from Barrow's groping for different tactics to increase his audiences' understanding.

Barrow began these lectures with the assumption that "there is really no Quantity in Nature different from what is called *Magnitude* or *continued Quantity*."[68] Magnitude alone was to be the object of mathematics, and thus all mathematics was for him subsumed under geometry. Arithmetic had no independent standing. Moreover, all "mixed mathematics", or natural science, dealt with objects that have magnitude and was therefore dependent upon geometry. He stated, "...those which are called *Mixed* or

---

[65] Barrow, *Geometrical Lectures*, p. 176.
[66] Barrow, *Geometrical Lectures*, p. 176.
[67] Barrow, *The Usefulness of*. . ., pp. 234-5.
[68] Barrow, *The Usefulness of*. . ., p.20

Concrete Mathematical Sciences are rather so many Examples only of Geometry . . . ."[69]

He also said,

> For Magnitude is the common Affection of all physical Things, it is interwoven in the Nature of Bodies, blended with all corporeal Accidents, and well nigh bears the principle Part of the Production of every natural Effect.

Indeed, all of physics was part of Geometry for Barrow.

Barrow broke with both the classical division of mathematics into arithmetic and geometry and the Aristotelian distinction between pure and applied mathematics. The latter division refers to the separation between pure mathematics, which deals with the intrinsic properties of number and magnitude, and applied mathematics, which includes such things as music and astronomy. Although he was quite familiar with the traditional opinions, he disregarded them in order to reformulate the foundation of mathematics.[70] He wanted even "mixed" mathematics to be considered a branch of geometry.

Barrow's opinion of algebra (which was also referred to as analysis and included techniques which we would consider 'pre-calculus') is often quoted, but historians often quote it out of context by not taking into account his other statements on this subject:

> Because indeed *Analysis*, understood as intimating something distinct from the Rules and Propositions of Geometry and *Arithmetic*, seems to belong no more to *Mathematics* than to *Physics, Ethics* or any other Science. For this is only a Part or Species of *Logic*, or a certain Manner of using Reason in the Solution of Questions . . . . Wherefore it is not a Part or Species of, but rather an Instrument subservient to the Mathematics. . . .[71]

Thus for Barrow, algebra was not a part of arithmetic, or even a part of mathematics at all. This view, however, was not a product of hostility towards modern methods in mathematics, as historians have often assumed. He greatly valued algebra as a method of discovery; he said, " The new Method of *Analysis*, chiefly cultivated by *Vieta* and *Cartesius*, [is] almost equal to every soluble question."[72] Moreover, the mathematical examples given above make it clear that Barrow was capable of using algebraic methods.

Barrow's conception of number, for all his respect for the classical traditions, was decidedly novel when compared with Greek ideas. He was convinced that number was "continued quantity" and that therefore arithmetic and geometry dealt with the same subject. Although he knew he was flouting the classical distinction between arithmetic and geometry, he claimed the mantle of a restorer, stating:

> I will rather restore it into its lawful Place, as being removed out of its proper seat, and ingraff and unite it again into its native Geometry, the Stock from whence it has been plucked...[73]

But unless, like Stevin, he had an undiscussed belief in a pre-Greek "golden age," the use of the word restore is odd. He makes it quite clear that he knew that Aristotle thought arithmetic was more general than geometry because a unit, unlike a point, does not require position.

---

[69] Barrow, *The Usefulness of. . .*, p.27.
[70] Barrow, *The Usefulness of. . .*, p.16.
[71] Barrow, *The Usefulness of...*, p.28.
[72] Barrow, *The Usefulness of...*, p. 243.
[73] Barrow, *The Usefulness of..*, p.29

Barrow was partially responding to mathematicians such as Wallis who claimed that numbers are more abstract than geometric magnitude when he separated numbers into two kinds. First, mathematical numbers are used by mathematicians to express and compare together the dimensions of homogeneous magnitudes. Second, transcendental or metaphysical numbers are "indifferently" attributed to all things, even if they are heterogeneous.[74] The kind of abstract number that Wallis referred to he considered to be transcendental and not mathematical, thus for Barrow it was outside the bounds of discussion. The proper object of study was the mathematical number; as Barrow put it,

> a *Mathematical Number* has no Existence proper to itself, and really distinct from the Magnitude it denominates, but is only a kind of *Note* or *Sign* of Magnitude considered after a certain Manner.[75]

and also

> For I am convinced that Number really differs nothing from what is called Continued Quantity, but is only formed to express and declare it; and consequently that Arithmetic and Geometry are not conversant about different Matters, but do both equally demonstrate Properties common to one and the same Subject; and very many, and very great Improvements will appear to be derived from hence upon the Republic of Mathematics.

Numbers were to be thought of as mere symbols for magnitude. Arithmetic would therefore become another way of dealing with magnitude that used symbols instead of figures and computation instead of construction.

Integers, then, would become the symbol of magnitude that was composed of equal parts. So if the magnitude *A* consisted of six equal parts, then *A* would be called six. Fractions were to be thought of as a sort of ratio, or comparison, between magnitudes composed of the same type of parts. Lastly, surd numbers, or irrationals are "Notes shewing a Magnitude to be any way a *Mean Proportion* between some assumed homogeneous Magnitude . . . ."[76]

Indeed, Barrow was quite distressed about the exclusion of surds from the realm of numbers and about mathematicians referring to surds as being irrational, irregular, or inexplicable. Removing surds to the domain of algebra Barrow found particularly offensive because surds were arithmetic's "noblest and most profitable member" because when one is measuring or comparing magnitudes the results are more often irrational than rational. He believed that surds are necessary and thus have the same right to be considered numbers as integers and fractions.[77] Of course, if numbers are the notes or signs of magnitudes then surd, or irrational, numbers would become less of a problem as no mathematician in this period denied the existence of irrational magnitudes.

Barrow abandoned the ancient distinction between discrete number and continuous magnitude by making numbers mere symbols of magnitude. Another question is how he viewed the composition of continuous magnitude. Did he accept the classical dictum

---

[74] Barrow, *The Usefulness of*..., p.37.
[75] Barrow, *The Usefulness of*. . ., p.41.
[76] Barrow, *The Usefulness of*. . ., pp.43-44.
[77] Barrow, *The Usefulness of*. . ., p.44.

of infinite divisibility?[78] It is simple to ascertain Barrow's conception of magnitude because he wrote directly on foundational subjects. In his IX lecture, titled "Of the Termination, Extension, Composition and Divisibility of Magnitudes", he discusses the controversy about whether magnitude is composed of indivisible parts or is infinitely divisible. A full exposition of both ancient and modern opinions was given because he felt that this was a very difficult subject. Consideration was given to the arguments of the ancient atomists such as Epicurus and Lucretius as well as modern atomists such as Galileo. But in the end, Barrow concurs with the classic conception of magnitude being infinitely divisible. He says this idea "answers well to the Ideas of Men; and that most of the excellent Philosophers have supported and defended it."[79] Moreover, he claimed that "we can have no possible Conception of any Magnitude, but as compounded of *Parts*, nor of these *Parts* but as some how extended and consequently made up also of *Parts*."[80] He agreed with Aristotle that there could be no "least-part" of magnitude and he also cited Descartes' demonstration of the infinite divisibility of magnitude.

Barrow, however, did not want to reject the method of indivisibles. The method itself was highly valued; he called it "that excellent *Method of Indivisibles*, the most fruitful Mother of new Inventions in Geometry..."[81] Indeed, he thought so highly of the method that he said that it "never sufficiently can be praised."[82] But although he admired the technique, he felt its results could be restated rigorously through the method of exhaustion. He also realised the problems with utilising expressions such as "all these lines are equal to such and such a rectangle." [83] He believed that such remarks should be rephrased. For example, "all the lines" could be replaced by a sum of parallelograms "of a rather small . . . and inconsiderable height." At least if one insisted on referring to lines, he felt that one must view them as an infinite collection. In general, Barrow shared the Aristotelian belief in the infinite divisibility of magnitude, but he accepted the method of indivisibles.

Barrow was one of only a handful of scholars who discussed other subjects that were considered continua in the seventeenth century, such as space and time. He explored these topics in detail as many curves in this period were generated by motion. As we have seen, in his *Geometrical Lectures* he embedded his discussion of the properties of curves in the broader context of the generation of magnitude. A consideration of motion led Barrow to the topic of time. He believed that there exists a "great Affinity" between space and time. Space, for him, had the same relation to magnitude as time does to motion; time was considered to be in some sense the space of motion. Barrow's ideas about the nature of space and time were, like his foundation for numbers, counter to accepted Aristotelian principles.

Barrow's interest in space stemmed from the question of whether space is distinct from magnitude, as an important attribute of geometrical magnitude is that it occupies

[78] See chapter one for a summary of the traditional concept.
[79] Barrow, *The Usefulness of*..., p.152.
[80] Barrow, *The Usefulness of*..., p.150.
[81] Barrow, *The Usefulness of*..., p.186.
[82] Barrow, *The Usefulness of*..., p.243.
[83] Barrow, *The Usefulness of*..., p.198.

space. When we examine the idea of space we seem to define it by its extension and indefinite capacity, which were also considered to be the properties of magnitude. He realised that both ancients such as Aristotle and moderns such as Descartes believed that there was no space separate from magnitude, or indeed from matter. He could not agree. First, he supposed that matter could be finite, and since he believed that "God is infinite in Essence," he thought that a "space" beyond matter and extension was necessary in order not to place limitations on God. Second, he believed that God could, if he wished, create worlds beyond this world, thus God extends beyond matter and magnitude: space was assumed to be the capacity or possibility of magnitude. He concluded that "*Space* is a thing really distinct from Magnitude."[84] He also thought that "space is not any thing actually existent, and actually different from Quantity. . . .Space is nothing else but the *mere Power, Capacity . . . of Magnitude*."[85] His conception of time, however, was even less traditional than his ideas about space.

Time, for Aristotle, could not elapse without change. Once again Barrow did not accept the classical view, although he did see a relation between motion and time, stating "*Time* is in some sort the *Space* of *Motion*."[86] Instead, he posited the existence of absolute time, which is a quantity that one cannot measure except through some constant uniform motion such as the sun and the stars.[87] But time's existence did not depend on motion. Barrow's preoccupation with motion led him to also consider the composition of time.

He examined the continuous flow of a point and a moment or instant and decided that time was a continuum with one dimension. Time may be envisioned as the "trace of a continuously sliding instant, having some indivisibility by virtue of the instant,"[88] but the infinite divisibility of time does not imply that it is composed of instants, just as a line consists of points that do not compose it. He considered time to have many similarities to a line, and thought that time could be conveniently represented by a line. At times he seemed uninterested in this distinction and favoured using terminology that would let him get on with his work. He said that it did not matter whether one used the term instant or indefinite particle when referring to time just as it did not matter "whether we understand a line to be composed of innumerable points or of indefinitely small linelets."[89] The classical, Aristotelian view of time would not have allowed such manoeuvres in the name of brevity. When the traditional views conflicted with Barrow's techniques, he ignored them.

CONCLUSION

Barrow's conception of the relationship between numbers and magnitude was motivated by factors different from those that had previously affected the other scholars

[84] Barrow, *Geometrical Lectures*, p.176.
[85] Barrow, *The Usefulness of Mathematical* . . ., p. 176.
[86] Barrow, *The Usefulness of Mathematical* . . ., p. 184.
[87] Barrow, *Geometrical Lectures*, p.256.
[88] Mahoney,"Barrow's Mathematics," p.205.
[89] Mahoney,"Barrow's Mathematics," p.205.

previously discussed. For example, Barrow seldom worked with numbers when solving problems. His mathematical practice was geometric. He was also not interested in solving practical problems, even though he was willing to cite utility as a reason to study mathematics. Moreover, the advances made in algebra or mathematical symbolism did not influence his views about the relationship between number and magnitude. His reasons for offering a foundation for numbers appear to have been two-fold. First, he wanted to show that all mathematics, and indeed all science, should really be considered geometry.[90] In order to accomplish this, he needed to break down the ancient distinction between arithmetic and geometry. If numbers were really just notes or sign of magnitude, then arithmetic was merely geometry in a different guise. Second, he wanted to argue against mathematicians, such as Wallis, who wanted to give priority to arithmetic and who claimed that portions of mathematics that Barrow considered geometric, such as Euclid's second and fifth books, were really poorly done arithmetic.

Barrow's ideas about numbers and his concepts of space and time show us how far from traditional Greek ideas an early modern English mathematician could move, even without "external" influences. Although in many ways Barrow belonged both to the humanistic and Aristotelian traditions, and he at times tried to pass off his innovations as renovations, when Greek ideas conflicted with his goal of the unity of mathematics under geometry, he ignored them. To achieve the unification of mathematics under the banner of geometry he was willing to overthrow the ancient distinction between continuous magnitude and discrete numbers, the distinction between arithmetic and geometry, and the distinction between pure and mixed mathematics. In order to uphold his religious beliefs, he was also willing to create non-traditional concepts of space and time.

Barrow's foundation for number came at a price, accepting the unity of mathematics under geometry. But for some mathematicians, like Wallis, arithmetic and the new algebraic techniques seemed to have priority over geometry. Wallis' practice, unlike Barrow's, depended heavily upon numerical patterns and techniques. With Wallis, we will see a return to the acceptance of irrationals and fractions without foundations.

---

[90] Peter Dear, *Disciplines and Experience: The Mathematical Way in the Scientific Revolution* (Chicago: Chicago University Press, 1995), p. 223. Dear, on the other hand, believes that Barrow only maintained that mathematics was useful in all of physics.

CHAPTER 7

JOHN WALLIS

An investigation of John Wallis' mathematical practice and his views about the nature of number and its relationship to magnitude will demonstrate that there were several options available to justify the broadening of the number concept and the breakdown in the separation between number and abstract magnitude. One needs to accept the unity of mathematics under geometry in order to accept Barrow's foundation for number. But for Wallis, arithmetic and the new algebraic techniques had priority over geometry. Moreover, his practice depended a great deal upon numerical patterns and methods. Although Wallis did not state his opinions of continua concepts in as much detail as Barrow, his point of view will provide us with the ideas of a professional mathematician whose practice was mainly algebraic.

Wallis, unlike Barrow, was interested in a variety of practical pursuits. He deciphered Royalists' coded letters for the Parliamentarians during the Civil War.[1] His work on gravitation, presented to the Royal Society in 1674, discussed machines, strength of materials, and optimal loading weights for ships. Moreover, his concern with navigation led him to write many articles about the tides of the sea and their relationship with sailing vessels.[2] Indeed, in some ways Wallis had more in common with the scholars we studied in our earlier examples than with Barrow in that he mixed his mathematical practice with a keen interest in practical activities.

Wallis was also unlike Barrow in that he held geometry subordinate to arithmetic. He desired to free arithmetic from geometric representations. Figureless demonstrations were favoured for being more abstract. He stated "And I look upon this, as the great advantage of Algebra, that it manageth Proportions abstractly, and not as restrained to lines, Figures, or any particular Subject; yet so as to be applicable to any of the particulars as there is occasion."[3] For him, numbers were more abstract than magnitude. He worked with Euclid's Book V arithmetically, and he did not require homogeneity from his equations; he even stated in his *Mathesis Universalis* that "Book V of the *Elements*, although it is totally the theory of Ratios, is arithmetic rather than geometry.[4] Moreover, most of his important mathematical results seem to be due to a facility for

---

[1] J.F. Scott, "The Reverend John Wallis, F.R.S. (1616-1703)," *Notes and Records of the Royal Society of London* 15 (1960) p. 58.

[2] John Wallis, "An Essay of Dr. John Wallis, exhibiting his Hypothesis about the Flux and Reflux of the Sea," *Philosophical Transactions* 16 (1666):262-281, is a typical example.

[3] Wallis, *Treatise of Algebra*, p.399.

[4] Carl Boyer, *The History of the Calculus and its Conceptual Development*, p.179 and Chikara Sasaki, "The Acceptance of the Theory of Proportions in the Sixteenth and Seventeenth Centuries: Barrow's Reaction to the Analytic Mathematics," *Historia Scientiarum* 29 (1985), p. 94.

discerning numerical patterns. Like Briggs, Wallis had a wonderful intuition about numerical patterns that might have led him to also work on the binomial theorem for non-integer exponents. Wallis discussed the nature of numbers and their relationship to magnitude, although not in as much detail as Barrow. The portion of his account found in chapter ten of his *Mathesis universalis* (1657) was untypical for this period because he included a treatment of the integers that were usually taken as a self-evident quantity.[5] Although he utilised fractions and irrationals, he strove to adhere to the ancient terminology, and when he was forced to depart from it, he always supplied an extenuating justification.

Wallis will supply us with the point of view of a scholar working almost entirely within the new algebraic and analytic tradition and who was completely immersed in the new science. Unlike Barrow, he was entirely uninterested in geometry for its own sake. Even his work on conic sections was meant to show how treating the conics as plane curves, and using equations, improved on the "obscure and involved methods followed by Apollonius." He also made free use of the method of indivisibles, and at least at the early stages of his career, he admired the analytical methods of Descartes. But although Wallis was an adherent of the latest mathematical methods, with him we see a return to the confusion over the exact status of fractions and irrational numbers. Although he was willing to utilise them in his computations, he was uncertain as to their exact status as numbers.

This chapter will first briefly investigate Wallis' education and mathematical training. This will provide us with some idea of what influenced his particular opinions aside from the general ideas discussed in the first chapter. It will also give us an overview of what mathematics Wallis was familiar with at the beginning of his mathematical career. Secondly, we will explore portions of Wallis' actual mathematical practice through, for example, a discussion of his treatment of Book II of Euclid's *Elements* and his use of numerical techniques in the *Arithmetica Infinitorum*. This will give us a picture of Wallis' mathematical research that will also supply us with his motivation for his positions on the nature of numbers and mathematical continua. Finally, we will examine the sections of Wallis' writing that provide us with his specific ideas relating to these topics.

BACKGROUND AND EDUCATION

Wallis was born in 1616 at Ashford. His father was a minister who died while Wallis was still a child in 1622. A plague in that area caused his mother to send him to a grammar school in Tenterden, but by 1630 he was at the Felsted

[5] Whiteside, "Patterns," pp. 187-188.

John Wallis, by Gerard Soest

Figure 1.

School that was discussed in the previous chapter. Wallis gave a brief account of his studies in his autobiography; they included not only Latin and Greek, but Hebrew and a smattering of Logic as well.[6] Apparently, he did not learn any mathematics at all at Felsted.[7] Wallis' first exposure to mathematics took place during the Christmas holidays of 1631. His younger brother, who was in the process of learning a trade, had for the previous three months been learning to "cipher" and cast accounts. Wallis claimed that he learned everything his brother knew in ten days. He later said, "This

[6] Christoph Scriba, ed. "The Autobiography of John Wallis, F.R.S.," *Notes and Records. Royal Society of London* 25 (1970): 27.
[7] Grammar schools in England in this period did not provide even a rudimentary mathematical education. Such skills were taught outside by self-styled 'professors' of mathematics who were usually almanac-makers, astrologers, surveyors, gunners or gaugers. See Taylor, page 9 - 10.

was my first insight into *Mathematics* and all the *Teaching* I had."[8] He learned from his brother

> The Practical Parts of Common Arithmetick in Numeration, Addition, Subtraction, Multiplication, Division and Rule of Three (Direct and Inverse), the Role of Fellowship (with and without Time), the Role of False-Position, Rules of Practice, and Reduction of Coins, and some other things.[9]

The mathematics taught to those going into trade resembled the mathematics of the abacus school tradition. Wallis' early mathematical training was solely practical.

Wallis was sent to Emmanuel College, Cambridge, during the Christmas term of 1632, where he studied with Mr. Anthony Burgess, Mr. Thomas Horton, and Mr. Benjamin Whitchcot. Emmanuel College was well known for its religious radicalism; it was associated with the Presbyterian movement and it was a stronghold of Ramism.[10] Wallis was taught logic, ethics, physics, and metaphysics as well as anatomy, astronomy and geology. Although he was fond of mathematics, he did not study it formally at University, but merely worked on it "as a pleasing Diversion."[11] He claimed that this was due to the attitude towards mathematics at Cambridge at this time,

> For Mathematicks . . . were scarce looked upon as *Academical* studies, but rather *Mechanical*; as the business of *Traders, Merchants, Seamen, Carpenters, Surveyors of Lands*, or the like . . . . And amongst more than Two-hundred Students . . . I did not know of any Two (perhaps not any) who had more of Mathematics than I . . . [12]

It is difficult to ascertain whether this description of the state of mathematical education at Cambridge is accurate. Mordechai Feingold points out that Wallis' autobiography was written after he had been the Savilian Professor of Geometry for some fifty years.[13] Moreover, he was recalling a period when divinity, not mathematics, was his primary occupation. Perhaps students who wished seriously to study mathematics in this period could find the instruction they needed at Cambridge. As late as 1649, however, John Hall lamented the lack of demonstrations of mathematical theorems or instruments in universities.[14]

Wallis did not immediately take up a mathematical career. After finishing his M.A. and being ordained, he became Chaplain for a year to Sir Richard Darley and he then spent two years with Lady Vere. In 1644, he became one of the secretaries to the Assembly of Divines at Westminster.[15] During the Civil War, Wallis was encouraged

[8] Scriba, "The Autobiography of John Wallis", p. 27.
[9] J.F. Scott, *The Mathematical Work of John Wallis D.D., F.R.S. (1616-1703)* (London: Taylor and Francis, 1938), p. 4.
[10] Hugh Kearney, *Scholars and Gentlemen: Universities and Society in Pre-Industrial Britain, 1500-1700* (London: Faber and Faber, 1970), pp. 28 and 63.
[11] Scriba, "The Autobiography of John Wallis, F.R.S.," p.27.
[12] Scriba, "The Autobiography of John Wallis," p.27.
[13] Mordechai Feingold, *The Mathematicians' Apprenticeship: Science, Universities and Society in England, 1560-1640* (Cambridge: Cambridge University Press, 1984), p. 87.
[14] In *A Humble Motion to the Parliament of England Concerning the Advancement of Learning: and a Reformation of the Universities* as quoted in Richard Foster Jones' *Ancients and Moderns: A Study of the Background of the Battle of the Books* (St. Louis: Washington University Press, 1936), p. 102.
[15] Scriba, "The Autobiography of John Wallis," p.31.

to decipher Royalist codes for Parliament. In 1642 he was shown a letter written in cipher; it took him only two hours to decode it, and he continued with his decoding work until 1689. Decoding the dispatches of Charles I and Charles II for Parliament, Wallis proved to be so good at code breaking that he kept his position even after the Restoration. Wallis has been referred to as the English "father of cryptography".[16] In 1645, while living in London, he became part of a discussion group on the "New Philosophy or Experimental Philosophy."[17] They considered many subjects, a short sample includes: anatomy, geometry, navigation, statics, magnetics, "chymicks", mechanics and natural experiments. By 1647, Wallis' interest in mathematics had revived enough for him to begin a study of Oughtred's *Clavis Mathematicae*; he then believed he had found a new way to solve cubic equations.[18] After a Royalist was stripped of the post, he was appointed Savilian Professor of Geometry in 1649. The appointment seems to be due as much to his skill at code breaking as to his work with cubics.

After having learned from his brother "mercantile" mathematics, Wallis seems to have abandoned the subject until reading Oughtred's *Clavis Mathematicae.* Although he was soon to show mastery of Greek works, there is no indication that he was familiar with classical mathematics until he began his study of Euclid, Apollonius, and Archimedes in 1649. Sometime around 1650 Wallis read the works of Frans von Schooten, Viète, and Descartes, and by 1651 he had also read Torricelli's *Opera geometrica.*[19] In the space of a few years, from 1647 to the early 1650s, Wallis taught himself not only classical mathematics, but much of the modern works as well. Moreover, he remained deeply involved in the new experimental philosophy.

One problem with ascertaining the particular influences on Wallis' mathematics is that he had a long-lived quarrel with the French mathematical school. First, Fermat criticised Wallis' *Arithmetica Infinitorum* for its lack of rigour.[20] Next, they argued over Fermat's challenges to Wallis in number theory.[21] Although in his 1656 *Arithmetica Infintorum* Wallis had placed Descartes "with the greatest mathematicians of the age," by the time he published his *Treatise of Algebra,* in 1685, he compared Descartes unfavourably to Harriot.[22] In the historical section of his *Treatise*, Wallis consistently claimed to have been mainly influenced by his countrymen, and he also claimed priority for them in most instances. But he was undoubtably familiar with a

---

[16] David Eugine Smith, "John Wallis as a Cryptographer, "*Bulletin of the American Mathematical Society, 24* (1917): pp. 83-96; David Kahn, *The Codebrakers* (New York: Macmillian, 1967), pp. 166-169; See, especially, Peter Pesic, "Secrets, Symbols, and Systems: Parallels between Cryptanalysis and Algebra, 1580-1700" *Isis* 88 (1997), pp.674-692.

[17] Scriba, "The Autobiography of John Wallis," p.39.

[18] Christoph Scriba, *Studien zur Mathematik des John Wallis (1616 - 1703)* (Wiesbaden: Franz Steiner, 1966), p.3. Also see Wallis' *Treatise of Algebra,* page 121, where he explains that he afterwards found that his rule was "co-incident with Cardan's Rules."

[19] Scriba, *Studien*, pp. 3-5.

[20] Scott, pp.72-81.

[21] Christoph Scriba has explored Wallis' work with number theory in detail. See his *Studien zur Mathematik des John Wallis (1616-1703)*, especially chapter three "Zaheltheoretische Probleme aus unveröffentlichten Manuskripten."

[22] Scott, "John Wallis as a Historian of Mathematics," pp.336-337.

wide range of continental mathematics.

Johannis Wallis S. T. D.

Geometriæ Profeſſoris *SAVILIANI*,

in Celeberrima Academia OXONIENSI;

OPERA MATHEMATICA.

Volumen Primum.

OXONIÆ,

E THEATRO SHELDONIANO MDCXCV.

Figure 2. Title page of Opera Mathematica.

WALLIS' MATHEMATICAL PRACTICE

Wallis was interested in Book II of Euclid's *Elements*, and different versions of his translation appear in several of his works. His translations provide us with a counterpoint to Barrow's, and may help to reveal the differences between the "geometric" and "analytic" points of view.

A century before Wallis' and Barrow's versions of Book II were published, Ramus, a leading pedagogue, was among the first to suggest that algebra deserved greater importance.[23] He suggested that parts of the second and sixth book of Euclid's *Elements* were algebraic. Descartes and Ramus both held the position that, "algebra was only a vulgar (Descartes later said "barbaric") name for a sort of mathematics that the Greeks had used; traces of it were found, on the one hand, in Euclid and Pappus and, on the other hand, in Diophantus."[24] Thus Wallis' algebraic and numerical treatment of Book II belongs to an established mathematical tradition that Barrow was opposing when he made arithmetic another branch of geometry.

It is quite possible, however, that both Wallis and Barrow had read Henry Billingsley's 1570 English translation of Euclid's *Elements*, which was a very influential work. This edition's version of Book II contains a mixture of geometrical and arithmetical approaches. Proposition V of Book II begins with a traditional proof, which unlike either Wallis' or Barrow's renditions, uses geometrical constructions. Indeed, one can find a proof very much like it in Sir Thomas Heath's edition of Euclid's work.[25] But the author goes on to state, "This proposition is of great use in Algebra..."[26] He then gives an example of the proof using numbers, which might have influenced both Wallis and Barrow.[27] Niccolò Tartaglia's version, however, which they were also both familiar with, did not use numerical examples. It utilised only geometrical constructions.[28] Neither scholar followed this model.

In chapter xxii of his *Mathesis Universalis*, published in 1657, Wallis included an algebraic presentation of the propositions given in the Second Book of Euclid's *Elements*. *Mathesis Universalis* was a somewhat elementary work, consisting perhaps of the contents of his professorial lectures. In the beginning chapters, he discussed the basic operations of arithmetic and although his notation is clearer than earlier works, the content is not new.[29] He also gave a history of the development of arithmetical notation from classical times. Many operations were performed as many as five times using different notations. Wallis then declared the disadvantages of denoting algebraic powers by geometrical dimensions; he preferred to use arithmetical degrees.[30] He followed this scheme in his version of Euclid.

Wallis clearly supported the view that Book II was "arithematical" in nature;[31] he

---

[23] Michael S. Mahoney, "The Beginning of Algebraic Thought in the Seventeenth Century," in *Descartes Philosophy, Mathematics and Physics*, ed. Stephen Gaukroger (New Jersey: Barnes and Noble, 1980), p. 148.

[24] Mahoney, "The Beginning of Algebraic Thought," p. 148.

[25] Compare folilo 57 of *Euclides The Elements of Geometrie*, trans. by H Billingsley, (London, 1570) with Heath's *The Thirteen Books of The Elements* (New York: Dover, 1956), Volume I, p. 382.

[26] Billingsley, *Euclides The Elements of Geometrie*, fo. 57.

[27] The example is, "Take any number as *20*, divide into *2* equal partes *10* and *10* and then into two unequal parts *13* and *7*..," the example, although numerical, is given in words - not equations.

[28] Niccolò Tartaglia's, *Euclides Solo introdttore della scientie mathematice: diligentemente reassettato, et alla integrita ridotto per il degno prefessure di tal scientie*, (Vinegina, 1543) fo. xxxii.

[29] Wallis' discussion of numbers will be examined in the section of the chapter describing his continua concepts.

[30] At this point in his career, Wallis still respected Descartes. He was especially enthusiastic about Descartes' exponential notation. See John Wallis, *Mathesis Universalis*, p. 71 in *Opera Mathematica*, Vol. 1. His treatment of Book II, however, utilised Oughtred's symbolism.

[31] Indeed, in his *A Defense of the Treatise of the Angle of Contact*, p. 99, he says, "we find Euclide's Geometry (good part of it) employed about Numbers."

also found "the nature of *Algebra* being more Abstract, [than Geometry]."[32] Upon first glance at his proof of Book II proposition 5, the difference between his "proof" and the standard version given in Heath leaps out. The proof is usually stated as "If a straight line be cut into equal and unequal segments, the rectangle contained by the unequal segments of the whole together with the square on the straight line between the points of section is equal to the square on the half."[33] Instead of a construction of rectangles and squares, he constructs nothing at all.[34] He gives the usual statement of the theorem.[35] But Wallis immediately restates the problem,

> That is, if $Z = 2S = A + E$ then it will be that $S + V = A$, $S - V = E$, (just as $A - S = V = S - E$.) It will be that $AE + Vq = Sq$. Or (in numbers) if $12 = 6 + 6 = 8 + 4$ just as $8 - 6 = 2 = 6 - 4$. It will be that $8 \cdot 4 + 2 \cdot 2 = 6 \cdot 6$.[36]

Then he added a few equations and numeric examples:

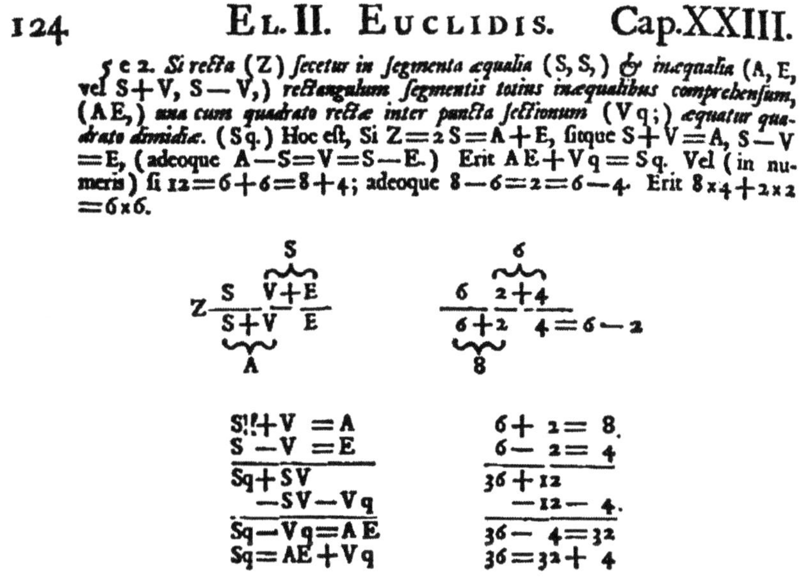

124 El. II. Euclidis. Cap. XXIII.

§ e 2. *Si recta* (Z) *secetur in segmenta æqualia* (S, S,) *& inæqualia* (A, E, vel S+V, S−V,) *rectangulum segmentis totius inæqualibus comprehensum,* (AE,) *una cum quadrato rectæ inter puncta sectionum* (Vq;) *æquatur quadrato dimidiæ.* (Sq.) Hoc est, Si Z=2S=A+E, sitque S+V=A, S−V=E, (adeoque A−S=V=S−E.) Erit AE+Vq=Sq. Vel (in numeris) si 12=6+6=8+4; adeoque 8−6=2=6−4. Erit 8×4+2×2=6×6.

[37]

Figure 3. Wallis's version of Euclid's Book 2 Proposition 5.

This ends Wallis' "proof." He does not justify these steps with reference to Euclid's definitions or theorems. It seems he believed that he had said enough. In this example,

---

[32] Wallis, *Treatise of Algebra*, p.90.
[33] Euclid, *Elements*, p. 382.
[34] Wallis, *Mathesis Universalis*, p. 124.
[35] Wallis, *Mathesis Universalis*, p. 124. If a straight line (*Z*) be cut into equal (*S,S*) and unequal (*A, E, or S + V, S - V*) segments, the rectangle contained by the unequal segments (*AE*) of the whole together with the square of the straight line (*Vq*) between the points of the sections is equal to the square on the half (*Sq*).
[36] Wallis, *Mathesis Universalis*, p. 124.
[37] Wallis, *Mathesis Universalis*, p. 124..

Wallis abandoned synthetic proof and geometrical constructions in favour of a heuristic approach.

In chapter xxvi of his *Treatise of Algebra*, Wallis again restated the theorems in Book II, this time without any diagrams. He said the theorems "... need no other Demonstrations than the actual Multiplication of the Terms...."[38] For example, proposition *1* was rendered:

> If $Z = A + E + I$; then is $BZ = BA + BE + BI$. (As will appear again by Multiplying both those equations into *B*.) And in like Manner (with little variation) for the following Propositions as themselves direct.[39]

He went on to give the other propositions as strings of equations. In this version of Book II, Wallis did not supply numerical examples. Both Wallis and Barrow used numerical examples in some problems though not in others.

Wallis' rendition of Book II of Euclid's *Elements* highlights certain aspects of his mathematical practice. In Wallis' version, the problems were treated as being purely arithmetical. He never utilised geometric constructions. Indeed, in other sections of *Mathesis Universalis*, he implied that geometry was subordinate to arithmetic.[40] Continuous magnitude in this period was usually treated geometrically, but Wallis felt numbers measured it. He suggested that in some respects number was antecedent to continuous magnitude. Wallis seems more comfortable when working with numbers, and rigour was not always his main concern.

Barrow, on the other hand, had viewed his use of algebra in his version of Book II purely as a convenience that could shorten his proofs. He moved back and forth between "algebraic" and "geometric" forms of proof, between equations, numerical examples and geometric constructions, while claiming that his presentation did not ". . . deviate in any particular from [Euclid's]."[41] Nothing about Book II convinced him that it was particularly "arithmetic." Thus, depending upon the mathematicians' point of view, Book II could be seen as arithmetic, or it could be viewed as showing that algebraic techniques were merely heuristic devises.

Wallis also worked with geometry using methods from different branches of mathematics in his work on conic sections. Indeed, he was the first to employ the Cartesian method in his *de sectionibus conics*, in that he believed that the conics should be regarded merely as plane curves and described by equations.[42] This work is, moreover, perhaps the first elementary textbook on conics that used the Cartesian methods. It is also the first text where Wallis presented his new method of indivisibles, where he attempted to free this method of criticism by considering a plane to consist of an infinite number of parallelograms having indefinitely small altitude. In the first part of the text Wallis discussed the fundamental propositions relating to the conics that had been investigated by Apollonius. The second part of the text showed how all these

[38] Wallis, *Treatise*, p.114.
[39] Wallis, *Treatise*, p.113.
[40] Wallis, *Mathesis Universalis*, p.69.
[41] See Chapter IV pp. 174-178.
[42] The date on the title page was 1655, but the work was issued as part two of *operum mathematicorum pars altera* (Oxford, 1656).

properties can be determined without reference to the cone.

In forty-four propositions, he developed conic theory from a basic definition as sections of a right cone using analytical equivalents of second degree equations:

$$e^2 = d(l - \frac{l}{t}d), p^2 = ld, h^2 = d(l = \frac{l}{t}d)$$

where *e, p, h* are the ordinates of the ellipse, parabola, and hyperbola respectively, *d* is the abscissa which was measured from the coordinate origin of the vertix of the conic, *l* is the latus rectum,[43] and *t* is the length of the main diameter conjugate to the ordinate.[44] This text is mainly of interest in that it shows that Wallis was familiar enough with Descartes' methods to employ them on the conics. It also shows how long it took for Descartes' techniques to be absorbed in England.

The mathematical problems that best display Wallis' mathematical characteristics have been dealt with in great detail in the secondary sources. But Wallis' skill at finding numerical patterns and intuition about continuity will best be brought out through these familiar examples.

Wallis' best known book, *Arithmetica Infinitorum*, published in 1656, describes his work on the quadrature of the circle. Although the techniques he used have been described in detail in J.F. Scott's *The Mathematical Work of John Wallis D.D., F.R.S. (1616 - 1703)*, a brief overview of sections of the material will provide examples of Wallis' mathematical style.

Wallis' method in this work is nonrigorously based upon what seems to be his strong feeling for pattern, and although many of the methods are similar, Wallis was not familiar with Briggs' numerical methods.[45] His techniques, however, might be related to his work on codes during the English Civil War. Evidently the whole layout on the printed page corresponds with the natural way one sets out a message for decoding. The codes used in the Civil War were numerical, and there were easily recognisable frequency patterns occurring among the various number-sets used.[46] Many of the same skills one would develop while decoding would be useful for Wallis's style of numerical techniques.

In modern terms, the main focus of the *Arithmetica Infinitorum* was Wallis' attempt to evaluate the expression $\int (1 - x^2)^{1/2}\, dx$ between zero and one. He relied heavily on his "method of induction": one looks at a certain number of individual cases, observes the emerging ratios, and compares these with one another so that a universal proposition may be established. For example, in Proposition *1* he observed,

---

[43] The modern equation for the ellipse, for example, refers to the vertex as origin, i.e. $y^2 = kx\,(a - x)$. Putting $ka = p$, one obtains $y^2 = px - (p/a)x^2$, p is the latus rectum.

[44] Whiteside, "Patterns," p.295. Wallis, however, did utilise Descartes' notation, together with some of the symbols of Oughtred, but his explanations are still mainly verbal. Yet Scott, *Mathematical Works*, p. 24, quotes Hobbes as saying that this work "is so covered with the scab of symbols, that I had not patience to examine whether it were well or ill demonstrated."

[45] Whiteside, "Patterns," p. 236.

[46] Whiteside, "Patterns," p. 236.

$$\frac{0+1}{1+1}=\frac{1}{2}$$
$$\frac{0+1+2}{2+2+2}=\frac{1}{2}$$
$$\frac{0+1+2+3}{3+3+3+3}=\frac{1}{2}$$

so he concluded $(0 + 1 + 2 + \ldots + n) / (n + n + n + \ldots n) = 1/2$.[47]

He used this proposition to show how the area of a triangle could be easily determined. In Proposition *3* he says that a triangle is made up of an infinite number of parallel lines arithmetically proportional, starting from zero. Thus the sum of the lines making up the triangle is *1/2* the sum of those making up the rectangle of the same base and altitude because the ratio may be written as

$(0 + 1 + 2 + \ldots + n) / (n + n + n + \ldots n)$.[48]

In Proposition *19*, Wallis used his Method of Induction to discover the sum of

$$(0^2 + 1^2 + 2^2 + \ldots + n^2) / (n^2 + n^2 + n^2 + \ldots + n^2).$$

He noted

$$(0^2 + 1^2 + 2^2 + \ldots + n^2) / (n^2 + n^2 + n^2 + \ldots + n^2) = 1/3 + 1/6n,$$

through observing the fractions given below.[49] He realised that although the required ratio was over *1/3*,

$$\frac{0+1}{1+1}=\frac{1}{2}=\frac{1}{3}+\frac{1}{6}$$
$$\frac{0+1+4}{4+4+4}=\frac{5}{12}=\frac{1}{3}+\frac{1}{6}$$
$$\frac{0+1+4+9}{9+9+9+9}=\frac{14}{36}=\frac{1}{3}+\frac{1}{18}$$

the excess diminished as the number of terms in the series increased. Therefore, if the

[47] Wallis, *Opera Mathematica*, p. 365.
[48] Wallis, *Opera Mathematica*, p.366.
[49] Wallis, *Opera Mathematica*, p.373.

number of terms were infinite, he believed the ratio sought would be *1/3*. He then used this proposition to show that the volume of a cone is *1/3* of the circumscribing cylinder.

$$\frac{0+1}{1+1}=\frac{1}{2}=\frac{1}{4}+\frac{1}{4}$$
$$\frac{0+1+8}{8+8+8}=\frac{9}{24}=\frac{1}{4}+\frac{1}{8}$$
$$\frac{0+1+8+27}{27+27+27+27}=\frac{26}{108}=\frac{1}{4}+\frac{1}{2}$$

In a similar fashion, Proposition xxxix explores the case of a series of cubes of numbers which were arithmetically proportional. He showed that, in general,

$$(0^3+1^3+2^3+\ldots+n^3)/(n^3+n^3+n^3+\ldots+n^3)=1/4+1/4n.$$ [50]

PROP. I.

Lemma.

SI proponatur ſeries Quantitatum *Arithmetice-proportionalium* (ſive juxta naturalem numerorum conſecutionem) continue creſcentium, à puncto vel o (ciphra, ſeu nihilo) inchoatarum, (puta ut o, 1, 2, 3, 4, &c.) propoſitum ſit inquirere, quam habeat rationem earum omnium aggregatum, ad aggregatum totidem maximæ æqualium.

Simpliciſſimus inveſtigandi modus, in hoc & ſequentibus aliquot Problematis, eſt, rem ipſam aliquouſque præſtare, & rationes prodeuntes obſervare atque invicem comparare; ut inductione tandem univerſalis propoſitio innoteſcat.

Eſt igitur, exempli gratia, $\frac{0+1}{1+1}=\frac{1}{2}$ $\quad \frac{0+1+2=3}{2+2+2=6}=\frac{1}{2}$.

$\frac{0+1+2+3=6}{3+3+3+3=12}=\frac{1}{2}$. $\quad \frac{0+1+2+3+4=10}{4+4+4+4+4=20}=\frac{1}{2}$.

$\frac{0+1+2+3+4+5=15}{5+5+5+5+5+5=30}=\frac{1}{2}$ $\quad \frac{0+1+2+3+4+5+6=21}{6+6+6+6+6+6+6=42}=\frac{1}{2}$.

Et pari modo, quantumlibet progrediamur, prodibit ſemper ratio ſubdupla. Adeoque ——

Figure 4. Wallis's "proof".

[50] Wallis, *Opera Mathematica*, p.382.

By the Method of Induction, as in the other cases, it is clear that this ratio for an infinite number of terms becomes *1/4*. From these results he generalised, in Proposition *44*, that if one substitutes their *4*th, *5*th, *6*th and so forth powers, the required ratios will be *1/5, 1/6, 1/7*, and so forth. In an effort to expand on the generality of his results, Wallis investigated the cases of negative and fractional indices.

He began by asking what the value of n must be for the ratio to be *1:1*; he decided it must be zero. So the series whose corresponding ratio is *1:1* is

$$0^0 + 1^0 + 2^0 + 3^0 + \ldots.$$

Moreover, he believed his rule would still be legitimate if one started with any series (with n integer, say *2* or *4*) and formed from it new series whose terms are the squares or cubes of the initial series. Also, if one starts with a series such as $0 + 1 + 8 + 27 + \ldots$ or $(0^3 + 1^3 + 2^3 + 3^3 + \ldots)$ whose corresponding ratio is *1:4*, one can form from it other series such as $0 + 1 + 2 + 3 + \ldots$ whose terms are the cube roots of the original series and $0 + 1 + 4 + 9 + \ldots$, which is the squares of the cube roots. The two derived series have a corresponding ratio of *2* and *3* that may be regarded as being obtained by the interpolation of the arithmetic means between *1* and *4*. Wallis felt that this principle of continuity, or interpolation was the key to fractional indices.

In Proposition 53, Wallis considered the series

$$\sqrt{0} + \sqrt{1} + \sqrt{2} + \sqrt{3} + \ldots$$

whose terms are the square roots of the series

$$0 + 1 + 2 + 3 + \ldots.$$

He deduced that the corresponding ratio of this series should be the arithmetic mean of the consequents of the two series,

$$0^0 + 1^0 + 2^0 + 3^0 + \ldots$$

$$0^1 + 1^1 + 2^1 + 3^1 + \ldots$$

whose ratios have the consequents *1* and *2*. So the series of square roots ought to have a corresponding ratio *1:1 1/2*.[51] Generalising further, he extended his argument to all positive rationals, and displayed his results in a table in Proposition *54*.[52] He explained,

> If an infinite series of quantities be understood, starting from *0*, and continually increasing according to the ratio of the square roots, cube roots, fourth roots, etc. of numbers arithmetically proportional. . . the ratio of the sum of the whole to the sum of an equal

---

[51] Wallis, *Opera Mathematica,* p.389.
[52] Wallis, *Opera Mathematica,* p.390.

number of terms each equal to the greatest will be as indicated in the following table.[53]

The table shows that a number of the form $^{a}\sqrt{r}$ can be expressed as a power of the form $r^{1/a}$, that the corresponding ratio of the series $0^n + 1^n + 2^n + 3^n + \ldots$ is $1/(n + 1)$ even if n is a fraction. He later showed that his result was also valid when n is a fraction of the type p/q.[54] Wallis had such faith in his "Method of Induction" that he assumed his results to be true even when n is irrational.

But in the end Wallis ran into difficulties with his method. According to his law of continuity, the ratio of the area under the curve to the area of its circumscribing rectangle should be $1 / (n + 1)$ even when n is negative. For the hyperbola $y = 1/x^2$, where $n = -½$, the area under he curve has the ratio to the circumscribing rectangle 1: -½ + 1, or 1:½. If $n = -1$, as in the hyperbola $y = 1/x$, this ratio becomes 1/0. He had, however, already decided in the *Conic Sections* that 1/0 was infinite. For Wallis, the ratio of a positive quantity to a negative quantity had no meaning. In the end he decided that such ratios were to be considered greater than infinite.[55]

The above examples demonstrate several of Wallis' mathematical characteristics: his extraordinary skill in finding numerical patterns led him to rely upon his "Method of Induction" and principle of continuity. Once he had made his discoveries using numerical methods, he did not hesitate to apply his results to geometric problems. He was also comfortable using algebraic notation. He said,

> For though such choice of Notes do not at all influence the Demonstration, yet doth it assist the Fancy and Memory, which would otherwise be in danger of being confounded . . .[56]

Additionally, his extended discussion of arithmetical notation from antiquity, along with his calling attention to the disadvantages of denoting algebraic powers by geometrical dimensions, helped to place the new notations on a firmer footing.[57]

THE COMPOSITION OF CONTINUA

Wallis, in the second chapter of his *Mathesis universalis*, adhered to the classical concepts by introducing the unit as "principium numeri" and "numerus" as "unitatum multitudo," or number as a multitude of units.[58] Perhaps he did this, as Jacob Klein claims, merely to save the definitions, as he went on to remark, "since number is commonly defined as a multitude of units . . ."[59] Also, he went on in chapter four to discuss at great length whether the unit was a number, referring to both Stevin and Descartes. In the end, he decided that one was a number.[60] For Wallis, like Stevin, the true principle of number was "nought," which he considered to be the analogue of geometric points.

---

[53] Wallis, *Opera Mathematica,* p.390, as translated in Scott, page 38.
[54] Wallis, *Opera Mathematica,* p.392.
[55] Scott, *Mathematical Work,* pp. 42-46. Scott provides a detailed examination of Wallis' difficulties.
[56] Wallis, *Treatise,* p.68.
[57] Scott, *Mathematical Work,* pp. 66-67.
[58] Wallis, *Opera Mathematica,* p. 20.
[59] Klein, *Greek Mathematical Thought,* p.211.
[60] Wallis, *Opera Mathematica,* pp. 24-27.

CAP. IV.

Unitatis & Numeri definitiones. Quo sensu Unum negatur esse numerus. Tam Unum quam Nullum, vario sensu, Principium Numeri. Unum est Numerus minimus. Nullum non est numerus. Numeri fracti, non sunt veri numeri.

Arithmetica derivatio & Definitio.
Unitatis & Numeri definitiones.
An Unitas sit Numerus.

ARithmetica, ἀπὸ τοῦ ἀριθμοῦ dicta, ab ipso objecto nomen sortita est. Est enim *Scientia numeri, quatenus est numerabilis*, vel *scientia bene numerandi*, ut supra declaratum est.

Numeri vero principium *Unitas* dicitur.

Unitas autem est, *secundum quam unumquodque Unum dicitur*: *Numerus autem ex unitatibus composita multitudo.* Sic enim Euclides, *lib. 7. de. 1, 2.* Μονάς ἐστι, καθ' ἣν ἕκαστον τῶν ὄντων ἓν λέγεται. Ἀριθμὸς δὲ, τὸ ἐκ μονάδων συγκείμενον πλῆθος.

Ex his autem Unitatis & Numeri definitionibus, plurimi arbitrantur Unitatem numerorum catalogo non accensendam, adeoque numerum minimum Binarium esse; quanquam illud quidem Euclidem uspiam dixisse non memini.

Mirror tamen eousque hanc sententiam invaluisse, cum non minus Unitas, quam binarius numerus, aut quivis alius, respondeat quaestioni *Quot sunt*.

Hac autem de re, erat mihi quidem in animo, litem neutiquam movere; cum illa controversia Logica potius seu Metaphysica, quam Mathematica, videri possit; saltem obiter monuisse, non modo *Unum* sed & *Nullum*, apud Arithmeticos eodem omnino modo ac reliquos numeros tractari; adeoque apud illos (quicquid de Metaphysicis dicendum sit) vel pro Numeris, vel quasi numeris reputari. Operae tamen pretium fortassis erit paucis ostendere, cur Unitatem putem vere numerum esse, aliisque numeris annumerandam.

Nec huic forsan adversabitur Euclidis aliorumque sententia probe intellecta.

I

Cum

Figure 5. Wallis's definitions.

He claimed that he was not really disregarding the unanimous ancient/modern opinion viewing the unit as the element of number. Indeed, he claimed to be agreeing with the ancients that the purely "arithmetic" methods of algebra deserved priority because the objects of arithmetic are higher and more abstract than those of geometry.[61] Although he mentioned that the ancients had accepted as numbers only the positive integers, and even said that "Euclid speaks only of true numbers and properly so-called [meaning integers] which are composed of units," he convinced himself that he was not being untraditional. He still doubted whether fractions could be considered true numbers, and thought that they would perhaps be better described as "fragments of the unit." [62] He believed that he was merely correcting their mistaken beliefs about the principle of number.

In the tenth chapter, he gave the now familiar treatment of the origin of the integers and operations on them. He ordered the integers by $n < n + 1$ and he allowed this to be extended to negative integers by letting $x = -a$ for $x + a = 0$. He then described the

[61] Klein, *Greek Mathematical Thought*, pp. 214-215 for a detailed discussion of this point.
[62] Klein, *Greek Mathematical Thought*, p. 220.

operations and rules.[63]

Wallis accepted the traditional division of pure mathematics into two disciplines: arithmetic and geometry. He stated, "geometry is the science of magnitude insofar as it is measurable, and arithmetic the science of number insofar as it is countable." [64] He considered algebra to be a part of arithmetic, but he did not regard algebra as a modern subject. In his *Treatise of Algebra, Both Historical and Practical,* he said algebra "was in use of old among the *Grecians*; but studiously concealed as a Great Secret." [65] He firmly believed that the Greeks possessed a secret "Art of Invention" used to discover the propositions which they then demonstrated in other ways. Indeed, he followed Viète in believing algebra to be a rediscovery of ancient principles; it was merely improved by the new symbolism. For Wallis, a preference for algebra was only favouring one branch of classical mathematics over another, not a radical break with tradition.

Wallis was following traditional Aristotelian dictates when he claimed that arithmetic is more abstract and universal than geometry.[66] Aristotle had argued that points implied position and units do not, therefore units were simpler than points. Wallis agreed that arithmetic is more general than geometry because of the wholly abstract nature of numbers. He said the assertion that $2 + 3 = 5$ is more general than a statement which merely applied to geometrical objects because,

> For two angels and three angels also make five angels. But the same argument holds for all operations, whether arithmetical or specially algebraic, which proceed from more general principles rather than being restricted to geometrical measures.[67]

The assertion that arithmetic, including algebra, is more universal and abstract than geometry was the sort of idea that Barrow was resisting with his definition of numbers as notes or signs of magnitudes. Barrow claimed his geometrisation was restoring the classical tradition, but Wallis denied that his method was in conflict with it.

The dissimilarity in Wallis' and Barrow's views on arithmetic and algebra can be explained, in part, by their conflicting opinions about mathematical objects, such as numbers. Wallis' views on the nature of numbers were not presented in one place in a systematic fashion; indeed they sometimes conflict with each other. In Chapter LXVI of his *Treatise of Algebra*, during a discussion of "Imaginary Quantities", he implies that numbers are positive integers only. Negative quantities are necessary, but they are not numbers. He says,

> But it is also Impossible, that any Quantity (though it not a supposed square) can be *Negative*. Since it is not possible that any *Magnitude* can be *Less than Nothing*, or any *Number Fewer than None*. [68]

Although negative quantities are not, strictly speaking, numbers, Wallis went on to explain how they may be understood.

Given the equivalent of a number line, with a reference position marked off as *A* and

---

[63] Whiteside, "Patterns," pp. 187-188.
[64] Mahoney, "Barrow's Mathematics," p. 189.
[65] Wallis, *Treatise,* introduction.
[66] Barrow, p.47. It is interesting that Barrow brings up Wallis' ideas on this issue.
[67] Mahoney, "The Beginning of Algebraic Thought," p.190.
[68] Wallis, *Treatise*, p.265.

with unit markers on either side of *A*, one may explicate the meaning of say *-3*, by letting a man walk *5* units to *B*, and then "retreat" *8* units to *D*, then *5 - 8 = -3*. Wallis claims that *-3* designates the point *D*. Thus negative quantities are developed in terms of an integer "number line" without being admitted as actual numbers.

Wallis discusses fractions in terms of division. Sometimes, he felt division to be impossible. For example, if *2* is to be divided into *3* equal parts, it is impossible because "the nature of Numbers doth not permit . . ." this type of operation. Incommensurable quantities were also not thought to be expressible in terms of numbers. However, he believes they could be approached as closely as one liked by continual approximation.[69] Two ways of dealing with surd quantities are discussed. He proposes that Stevin's method of decimal fractions was very useful in cases that were not in need of mathematical exactness. A series of approximations gave him decimal representations for such numbers as π and $\sqrt{2}$. One could also utilise infinite series through interpolation.

Wallis found only positive integers to be numbers in his *Treatise of Algebra*; negative numbers, fractions and surds were considered to be quantities. But in his *A Defense of the Treatise of the Angle of Contact*, published as an appendix to his work on algebra, he seems to broaden his definition. In Chapter V, "Concerning the Composition of Magnitudes," he discussed the classical belief that numbers were composed of undivided units. The "Greek Authors" had believed that arithmetic could only be applied to positive integers. The classical opinion was that numbers were multiples of indivisible units. Wallis, however, was willing to extend the field of arithmetic "not only to Fractions, but to Surds also."[70] He went as far as saying "It is hard to say, what bounds can be set to it [arithmetic]; or, to what it may not extend."[71] But what does this mean? By "Arithmetick" he certainly meant to include algebra, but he never clearly states that fractions and surds were numbers, merely that one could operate on them algebraically. Barrow's definition of number, in contrast, did not include these limitations.

Another question is how Wallis viewed the composition of continuous magnitude. Did he accept the classical dictum of infinite divisibility? Recall that Aristotle believed that lines, for example, were not composed of points, or any other indivisibles, as this would require the points to be immediately next to each other. Thus points would have to have parts in order for their extremities to touch, but this would contradict the "definition" of point. In fact, Aristotle defined continuous as "that which is divisible into divisibles which are infinitely divisible."[72] Wallis did not deal with such issues directly, but we may investigate his views by examining his work on indivisibles. He learned of Cavalieri's methods by 1653 through a study of Torricelli's works. Wallis employed some of the symbolism and equations for curves of analytic geometry as well as the method of indivisibles in both his *de sectionibus conics* and his later work focusing on the quadrature of the circle, *Arithmetica Infinitorum*, published in 1656.

---

[69] Wallis, *Treatise*, p.317.
[70] Wallis, *Treatise*, p.92.
[71] Wallis, *Treatise*, p.92.
[72] C. Jones, p.47.

Mathematicians in this period realised that there were problems with Cavalieri's technique of regarding a line as made up of an infinite number of points and surfaces to be made of an infinite number of lines. Wallis attempted to circumvent this difficulty in two ways.

First, he claimed that the method of indivisibles was "not, as to the substance of it, really different from the Method of Exhaustions."[73] It was merely a short cut, which could be rephrased in more rigorous terms. Thus, if one doubted a certain result, it was always possible to reformulate it in terms of the method of exhaustion. Second, instead of lines having no breath making up a surface, Wallis introduced "small surfaces (of such a length, but very narrow)", or infinitely small parallelograms.[74] He was not the first to improve Cavalieri's method in this fashion as Pascal and Roberval had made this suggestion earlier. This innovation does suggest, however, that he did not think a surface could be constructed out of lines with no breath, and it is therefore unlikely that he would have assumed that a line could be built up from points. Thus Wallis adhered to the typical belief in the infinite divisibility of magnitude, and gave no indication of holding any sort of "atomic" views.

Although Wallis was, in general, more interested than Barrow in heuristic analysis, and he was willing to utilise the latest methods, he responded sharply to any hint, such as those made by Fermat, that his methods were not rigorous. While admitting that he usually tried to "show a way of *Investigation* or finding out of things yet unknown: (Which the Ancients did studiously conceal)," he was still convinced that all of his work could be recast into a more rigorous format.[75] He felt it was mainly a question of brevity. Why take up so much time with rigorous methods when the results would turn out to be the same? But he did not feel that he was being untraditional.

CONCLUSION

Wallis struggled to incorporate the new symbolism and algebraic techniques into his work while simultaneously maintaining the traditional concepts of number and magnitude. Although, like Stevin, he considered the unit to be a number, he was doubtful as to fractions' and irrationals' status as numbers. He stated, "I admit that the numbers which are called 'broken' or 'fractions' are, as it were, in between 'one' and 'nought' ... [but they are] not so much a number as an index of the ratio numbers have to one another."[76]

But Wallis was willing to work with both fractions and irrationals. He considered them in some sense to be included under mathematical entities dealt with by arithmetic, or algebra. Klein, for example, argues that for Wallis, "number", considered as symbolically conceived ratio, became the object of arithmetic, or algebra.[77]

Wallis' work continued the movement towards conceiving numbers as symbols

[73] Wallis, *Treatise,* p.285.
[74] Wallis, *Treatise,* p.286.
[75] Wallis, *Treatise,* p.305.
[76] Klein, *Greek Mathematics*, p.220.
[77] Klein, *Greek Mathematics*, p.223.

insofar as he considered variables to be equally able to represent both numbers and magnitudes. Once again, what would be accepted as an appropriate value for the variable seems to have been heavily influenced by the type of problems a mathematician was trying to solve. Wallis was influenced, as were the early modern English algebraists, by the notion that algebraic symbols could represent both numbers and magnitudes, or any "mutually homogeneous quantities in continuous proportion."[78] Moreover, the type of problems he wanted to solve motivated him to accept fractions and irrationals as a type of mathematical object, if not as "true numbers so called." He attempted, in his *Arithmetica Infinitorum*, to evaluate the expression $\int (1 - x^2)^{1/2}\, dx$ between zero and one, using his "method of induction" and his "principle of continuity." These methods, however, required him to utilise both fractions and irrational numbers. Moreover, it is possible that his practical work with decoding suggested this approach to him.

Wallis' work shows that the question of the status of fractions and irrationals had not been completely resolved even in the later part of the seventeenth century. Barrow's solution of making numbers the notes or signs of magnitude and subsuming all of mathematics under geometry was not universally accepted. The traditional concept of number seems to have coexisted, after a fashion, with the utilisation of fractions and irrationals. Although they were not quite numbers "properly so called," they were extremely useful, and Wallis attempted to retain the classical terminology rather than the classical practice.

[78] Wallis, *Mathesis universalis*, p.57.

CHAPTER 8

CONCLUSION

The English and Scottish mathematicians we have examined had several options available for dealing with numbers and their relationship to magnitude. Although they were familiar with Aristotle's and Euclid's distinction between discrete number and continuous magnitude they might have simply ignored it, or declared it to be mistaken. Neither of these paths was taken. Napier, for example, when he discussed these matters at all, concurred with the traditional point of view for pure mathematics. For Napier, it was only in applied mathematics that numbers and magnitudes could take on new relationships, but no explicit rejection was made of Greek ideas, nor was any justification offered for departing from them. Barrow, on the other hand, pretended his renovations were merely "restoring" proper mathematical relationships, although he admitted that he was diverging from Aristotelian concepts. Thus despite the pressure from the changing nature of the problems and the development of new mathematical methods to broaden number concepts, outright rejection of the traditional framework was not chosen.

Another possible way of dealing with the tension between the requirements of problem solving, the new mathematical techniques, and appreciation of classical concerns was to expand the number concept while simultaneously separating the new entities from "true" numbers. This was the most popular method. Numbers were defined in the traditional way as being multiples of units, and then they were expanded by *ad hoc* extensions of "broken" and "surd" numbers.[1] This solution merely sidestepped the issue as it was never properly explained what the relationship was between the traditional definition of numbers as being multiples of units and these new numbers. On the one hand, Recorde, for example, dealt with numbers that could not be expressed as positive integers by giving them the special names "broken" or "surd" numbers. On the other hand, Recorde stated that "Unitie is of it self indivisible," and that unity is the origin of number, and "One is no nomber." However he never explained the relationship between his classical definition of numbers and these extensions of the number concept. In this way he was able to solve certain problems requiring non-integer solutions while at the same time not explicitly abandoning the classical framework.

---

[1]"Broken" meaning a fraction, or rational number that could not be expressed in term of integers, and "surd" meaning irrational.

Yet another possibility would have been to offer an alternative foundation that would justify the interchangeable use of numbers, symbols, and magnitude. This route was never explicitly taken. For instance, Harriot's unpublished manuscript challenged the Aristotelian notion of the composition of magnitude. He decided that a continuum is an aggregate of things in contact, or that continuity is composed primarily of indivisibles in contact with each other, but he never connected his revision of the concept of magnitude to ideas about number. Moreover, he wrote ambiguously in English at the end of the manuscript, "Much ado, about nothing. Great warres and no blowes. Who is the foole now?"[2] In addition, although there is evidence in his manuscripts that Harriot realised that not all solutions to equations could be written as integers (or even as fractions), his posthumously published *Praxis* did not address these issues.

Although Napier's works on logarithms implicitly challenged traditional assumptions by constructing a one-to-one correspondence between number and geometric magnitude, and by writing numbers as decimal expansions, he never offered an alternative foundation that would formally justify these moves. His use of kinematic techniques seems to indicate that he wished to supply an intuitive understanding of how numbers might be linked to magnitudes, but he never straightforwardly stated that number and magnitude were interchangeable. Briggs' development of Napier's logarithms did not even offer a kinematic framework; he treated numbers as though they were arbitrarily closely approximal by terminating decimals without supplying a justification of any kind.

When we turn to Barrow's and Wallis' work, the earliest attempt to construct some kind of framework for the broadened use of numbers and their new relationship to magnitude, it is apparent that the classical concepts were not completely abandoned. One the one hand, Barrow's conception of number was decidedly different from the Greek idea. He was convinced that number was "continued quantity," and therefore arithmetic and geometry dealt with the same subject. On the other hand, he claimed the mantle of a restorer, although he knew he was flouting the classical distinction between arithmetic and geometry. He rejected one classical view to save another. For example, Barrow wanted to argue against the position that Euclid's second and fifth books were really poorly done arithmetic.

In many ways Barrow belonged both to the humanistic and Aristotelian traditions. He tried to present his innovations as renovations, but his most cherished goal was of the unity of mathematics under geometry. To achieve the unification of mathematics under the banner of geometry, he was willing to discard the ancient distinction between continuous magnitude and discrete numbers, the distinction between arithmetic and geometry, and the distinction between pure and mixed mathematics. In order to uphold his religious beliefs, he was also willing to create non-traditional concepts of space and time. Here too he also attempted to present his ideas as a restoration of an ancient tradition.

---

[2] Hilary Gatti, *The Renaissance Drama of Knowledge: Giordano Bruno in England* (London: Routledge, 1989), p.65.

Wallis also modified, without explicitly abandoning, the classical concepts. He struggled to incorporate the new symbolism and algebraic techniques into his work while simultaneously attempting to maintain the traditional concepts of number and magnitude. Wallis considered the unit to be a number, but he was doubtful about the numerical status of fractions and irrationals. In some sense "broken" numbers, or fractions, seemed to him to be between "one" and "nought," but he viewed them less as numbers as an index of the ratio numbers have to one another. He was, however, willing to work both with fractions and irrationals. He considered them in some sense to be included under mathematical entities dealt with by arithmetic, or algebra. For Wallis, "number" considered as a symbolically conceived ratio, became the object of either arithmetic or of algebra. On the one hand, Wallis was willing to modify the Greek view by considering "one" to be a number. On the other hand, he was not comfortable in labelling fractions and irrationals "true numbers".

Looking at mathematical practice shows a great deal of tension between the scholars' concern with the classical dictates on the nature of number and its relationship to magnitude, the requirements of problem solving, and the evolving mathematical methods. Once algebraic techniques were utilised, however, they encouraged a certain interchangeability between symbols, numbers, and abstract magnitudes. Klein's supposition of the close relationship between an algebraic mode of thought and numbers being treated as symbolic, instead of as discrete collections counting definite objects, is sustained by English algebraic practice. Even the earliest work of Recorde, before the introduction of Viète's ideas, shows movement in the direction of number being perceived more symbolically. Recorde states: "And although there be many kindes of irrationall nombers, yet those figures that serve in Cossike nombers, bee the figures also of all irrational nombers. .."[3] In other words, the numbers that are found to be the value of the cossic numbers when the problem is solved might well be irrational. Recorde's work, especially when compared to arithmetics written in this period, shows that an interest in solving practical problems is not enough, in itself, to cause a brodening of number concepts. It requires the combined impact of both novel practical applications and algebraic notation. As we have seen, however, even with both these factors, Recorde did not relinquish the traditional definition of number. Although the algebra of Harriot and Oughtred became increasingly symbolic, and the number concept broadened to include the solutions to more complex equations, the traditional view of number was never explicitly rejected.

The adoption and development of logarithms also brings along with it certain tacit theoretical implications. Napier's work implies that number and magnitude are in some ways the same *kind* of thing. It is difficult to hold firmly to the idea of discrete number *versus* continuous magnitude when you have placed numbers and magnitude into a one-to-one correspondence and computed decimal expansions of numbers to the desired degree of accuracy. The logarithms also helped to spread the broadened number concept to a wider community. Later texts that deal with logarithms, such as Edmund Wingate's *Arithmetique Made easie* (1630), use the algebraist's manoeuvre of separate

[3]Recorde, *Whetstone*, Sig. Sii[r]

labels such as "broken" or "mixed" numbers. Wingate even defined numbers as "that by which everything is numbered," which allows numbers to be interpreted fairly broadly. Thus logarithms and algebra, two of the most important mathematical techniques developed in Britain in this period, both motivated the acceptance of a broadened number concept and helped to shape opinions about the relationship between number and magnitude, but their influence was not strong enough to cause an outright rejection of classical concepts.

Logarithms had an indirect influence on English and Scottish ideas about mathematical continua. They were so useful that they accustomed a large variety of people to working with decimal expansions and irrational numbers. Distinctions made by Euclid, such as the separate theories of proportion for magnitudes and numbers, were not discussed. Although no foundations were available for the real numbers, the utility of the logarithms encouraged this lack of proper foundations to be pushed aside. The impact of logarithms was similar to the influence of algebra. The use of algebra introduced problems whose solutions were irrational. As algebra became to be viewed as being extremely useful both mathematically and practically, it influenced its users to accept a wider variety of numerical solutions. The early algebra of Recorde exhibits concern with the status and labelling of numbers that were not positive integers. By the time of Harriot's work, however, it was merely accepted that sometimes solutions would be irrational. Similarly, the logarithms broadened the range of allowable solutions to problems. They helped to spread the technique of decimal expansions, and thus furthered the notion that numbers were not only discrete collections of units.

The changing nature of the problems being solved also influenced these developments. Some of the problems English and Scottish mathematicians were attempting to solve did not have integer solutions. The character of these "actual problems" changed over time: Recorde's were often taken from practical activities such as measuring or building; his commercial problems also frequently required fractional solutions. Oughtred's might be the expression of a quadratic equation that he believed captured the essence of a Greek geometric problem; Harriot was usually interested in mathematics that would further his research in subjects such as free fall and optics. Napier's and Briggs' problems were mainly focused on simplifying trigonometric calculations. Trigonometry was useful in astronomy, surveying, map making, and navigation and much time was devoted to calculating extensive tables of sines and tangents. Moreover, safe and fast navigation was of the utmost importance during this period in Britain, and trigonometry was vital to problems in this area.[4] The logarithms also made multiplication, division, and extraction of roots, which were considered to be difficult operations in this period, easier and more accurate. Moreover, we have seen in Briggs' work how quickly logarithms were adapted to a wide range of practical activities. Many of these practical activities necessitated the broadening of the number concept, especially the increasing use of fractions, however they were also not enough to cause an outright abandonment of the classical framework.

---

[4]Herman H. Goldstine, *A History of Numerical Analysis from the 16th through the 19th Century* (New York: Springer-Verlag, 1977), p.13.

Another point that becomes clear after examining English and Scottish mathematical practice is that, in some sense, it was perfectly clear to the scholars we studied how to *use* the broadened number concept in a variety of problems. They were not puzzled about how to perform arithmetical operations on fractions and irrationals. There was, in a way, a "family resemblance" between these new types of numbers and the integers. In other words, they had a relationship with the objects that had previously called numbers, and they could be operated on in a similar way. It was a matter of making a choice concerning whether "broken" and "surd" entities should be called numbers because of their similarities with the integers, or "true numbers," or whether they wished to draw some kind of definitive boundary between different types of entities. There seems to have been little question about whether these objects should be utilised, it was more a matter of what they should be called.

The confusion over how the broadened numbers should be labelled seems to be partially proportionate to how familiar the scholar in question was with the classical concept of number. Some of the mathematicians we have studied, for example Recorde and Wallis, were very familiar with the Greek concept of number; thus although they were able to operate with an extended number concept, they were not sure if it was a legitimate move to label these new objects as numbers. The scholars that show the least awareness of the traditional concept of numbers as collections of discrete units, such as Napier and Briggs, were also the least interested in applying separate labels to the new objects in order to separate them from "true numbers." Napier, in his work on algebra, as we have seen, merely said that the traditional labels should be used when dealing with "pure" mathematics; in applied mathematics he did not bother with this distinction. For the students of the practical mathematics texts, who were not steeped in an Aristotelian tradition, this concern over proper labelling would have been minimal. But most of the creative mathematicians in this period had a university education; they were concerned about the difference between their actual practice and the only available theoretical foundations.

Perhaps one reason for the hesitation in discarding Greek ideas was the lack of alternative formulations with which to replace them. Although Barrow claimed to be merely "restoring", his view of numbers and magnitude was genuinely different from classical concepts, and it entailed an acceptance of his view of mathematics. He wanted to show that all mathematics, and indeed all science, should really be considered geometric. In order to accomplish this goal, he needed to break down the ancient distinction between arithmetic and geometry. Arithmetic was merely geometry in a different guise, because numbers were really just notes or sign of magnitude. However for some mathematicians, like Wallis, arithmetic and the new algebraic techniques had priority over geometry, and Barrow's solution was unacceptable. Wallis simply returned to the pattern of accepting irrationals and fractions without foundations. Wallis' work shows that the numerical status of fractions and irrationals was not resolved in the later part of the seventeenth century. Barrow's solution of making numbers the notes or signs of magnitude and subsuming all of mathematics under geometry was never universally accepted. With Wallis the traditional concept of number seems to have coexisted, after a fashion, with the utilisation of fractions and irrationals.

Throughout early modern English and Scottish mathematical practice there was a tension between the usefulness of broadening the number concept and the lack of a proper foundation. The scholars were aware of the traditional theoretical issues, and they struggled to retain as much as possible of the classical framework as they were being called upon to answer questions that had not been asked before. The temptation to utilise the broadened number concept was great. The new numbers allowed them not only to meet the needs of customs officials, excise men, and gaugers but also to address the interests of contemporary mathematical practice in areas such as expanding the range of the binomial theorem and work with sequences and series. Moreover, the emerging idea that numbers, magnitudes, and symbols were interchangeable allowed problems from different domains to be treated in the same symbolic language which helped to uncover both the common underlying structures and the relationships between the structures.

The factors that influenced the gradual acceptance of fractions and irrationals and the breakdown in the classical distinction between number and abstract magnitude were the changing nature of the problems being solved, the increased use of symbolic notation, and evolving mathematical methods. All of these elements were necessary in order to overcome traditional dictates on the nature of number and magnitude, in practice, if not in theory. No consensus was reached in early modern England on these topics. The new symbolism and the new numbers, however, still became part of mathematical practice, if not of mathematical theory; they were too useful to be ignored.

REFERENCES

Aiton, Eric J. "Infintisimals and the Area Law." *Internationales Kepler - Symposium Weil der Stadt 1971.* Ed. F. Krafft, K. Meyer, and B. Sticker. Hildesheim: Dr. H.A. Gerstenberg.1973: 285-305.Alexander, Amir. "The Imperialist Space of Elizabethan Mathematics", *Studies in History and Philosophy of Science* **26** (1995):559-592.

Al-Khuwarizmi. *The Algebra of Mohammed Ben Musa.* Trans. Frederic Rosen. London: J.L. Cox, 1831.

Andersen, K. "Cavalieri's Method of Indivisibles" *AHES* 31 (1985): 291-367.

Andrews, Kenneth R. *Trade, Plunder and Settlement.* Cambridge, Cambridge University Press, 1984.

Aristotle. *Aristotle's Physics Book I and II.* Trans. W. Charlton. Oxford: Clarendon Press, 1970.

Aristotle's Physics Book III and IV. Trans. Edward Hussey. Oxford: Clarendon Press, 1983.

Metaphysics. Trans. Julia Annas. Oxford: Clarendon Press, 1976.

Archimedes. *The Works of Archimedes.* Ed. And trans. Thomas L. Heath. 2nd ed. Cambridge : Cambridge University Press, 1912.

Ashworth, William B. "The Sense of the Past in English Scientific Thought of the Early 17th Century: The Impact of the Historical Revolution". Diss. University of Wisconsin, 1975.

Asmis, Elizabeth. *Epicurus' Scientific Method.* Ithaca: Cornell University Press, 1984.

Aubrey, John. *Aubrey's Brief Lives.* Ed. Andrew Clark. 2 vols. Oxford, 1878.

Baker, Humfrey. *The Well Spring of Science.* London, 1568.

Baron, Margaret. "Napier, John." *Dictionary of Scientific Biography.* 609-613.

Barrow, Isaac. *The Usefulness of Mathematical Learning Explained and Demonstrated: Being Mathematical Lectures Read in the Publick Schools at the University of Cambridge.* Trans. Robert Kirby. London, 1734.

Euclide's Elements; The Whole Fifteen Books Compendiously Demonstrated. with Archimedes Theorems of Sphere and Cylinder, investigated by the Method of Indivisibles London: 1714.

Geometrical Lectures: Explaining the Generation, Nature and Properties of Curve Lines, Trans. Edmund Stone. London: 1735.

Isaaci Barrow Lectiones Mathematicae XXIII. London, 1685.

Billingsley, Henry. *Euclides The Elements of Geometrie.* London, 1570.

Blundeville, Thomas. *His Exercises, Containing sixe Treatises.* London, 1594.

Boas, Marie. *The Scientific Renaissance 1450-1630.* London: Collins, 1962.

Bos, H. J. M."On the representation of Curves in Descartes' *Géométrie.*" *AHES* 24 (1981): 295-388.

Redefining Geometrical Exactness: Descartes' Transformation of the Early Modern Concept of Construction. New York: Springer -Verlag, 2001.

Bostock, David. "Aristotle on Continuity in *Physics* VI." *Aristotle's Physics: A Collection of Essays.* Ed. Lindsay Johnson. Oxford: Clarendon Press, 1991. 180-212.

Boyer, Carl. *A History of Mathematics.* Princeton: Princeton University Press, 1985.

The History of the Calculus and its Conceptual Development. New York: Dover, 1959.

Briggs, Henry. *Arithmetica Logarithmica sive Logarithmorum Chiliades Triginta.* London, 1624.

Logarithmorum chilias prima. London, 1617.

British Library Add. MS. 6782, ff. 362-364v.

British Library Add. MS. 1628.4398, f. 132; 4412, ff. 1-6; 4419, ff. 92, 108-111, 143, 420; 4422, ff. 62-77, 80-84, 163b

British Library MSS Harleian 6796

Cajori, Florian. *A History of Mathematical Notations.* La Salle: Open Court Press, 1974.

William Oughtred: A Great Seventeenth-Century Teacher of Mathematics. Chicago: Open Court Publishing Co., 1916.

"Controversies between Wallis, Hobbes, and Barrow", *Mathematics Teacher* 2 (1929): 146-151.

"A Revaluation of Harriot's *Artis Analyticae Praxis*", *Isis* 11 (1928): 316-324.

"Algebra in Napier's Day and Alleged Prior Inventions of Logarithms". *Napier Tercentenary Memorial Volume.* Ed. Cargill Gilston Knott. London: Longmans, Green and Co., 1915. 93-109.
Caspar, Max. *Kepler.* Trans. C.D. Hellman. London: Abelard-Schuman, 1959.
Clark, Andrew. Ed. *'Brief Lives' chiefly of Contempories, set down by John Aubrey between the years 1669 and 1696*, 2 Vols. Oxford: 1898.
Clark, F.M. "New Light on Robert Recorde", *Isis* (1926), **8**, 50-70.
Coleman, D.C. *The Economy of England 1450-1750.* London: Oxford University Press, 1977.
Cormack, Lesley. *Charting an Empire: Geography at the English Universities, 1580-1620.* Chicago: University of Chicago Press, 1997.
Craven, Wesley Frank. *Dissolution of the Virginia Company: The Failure of a Colonial Experiment.* Gloucester, Mass. Peter Smith, 1964.
Davis, N. Z. "Sixteenth Century French Arithmetics and the Business Life." *J. Hist. of Ideas*, 21 (1960): 18-48.
Dear, Peter. "*Totius in verba*: Rhetoric and Authority in the Early Royal Society." *Isis* 76 (1985): 145-161.
Disciplines and Experience: The Mathematical Way in the Scientific Revolution. Chicago: Chicago University Press, 1995.
Descartes, Rene. *Discourse on Method, Optics, Geometry , and Meteorology*. Trans. Paul Olscamp. Indianapolis: Bobbs-Merrill Co., 1965.
Digges, Leonard and Thomas. *Stratioticos*. London, 1579.
Dijksterhuis, E.G. *Simon Stevin: Science in the Netherlands around 1600.* Hague: Martinus Nuhoff, 1970.
Drake, Stillman. "Euclid Book V from Eudoxus to Dedekind." in *History in Mathematics and Education: Proceedings of a Workshop held at University of Toronto, Canada, July-August 1983.* Paris: Belin, 1987. 52-64.
Galileo at Work: His Scientific Biography. Chicago: University of Chicago Press, 1978.
Easton, Joy. "Robert Recorde." *Dictionary of Scientific Biography.* Vol. XI. 338-340.
"The Early Editions of Robert Recorde's *Ground of Artes*." *Isis* 42 (1951): 208-218.
"A Tudor Euclid." *Scriptia mathematica* 27 (1966): 339-335.
Epicurus. *Letter to Herodotus*. Trans. George Strodach. *The Philosophy of Epicurus.* Northwestern University Press, 1963.
Euclid. *Elements*. Trans. T. Heath, 3 vol. New York: Dover, 1956.
Feinold, Mordechai. "Isaac Barrow: Divine, Scholar, Mathematician." *Before Newton: The Life and Times of Isaac Barrow*. Ed. Mordechai Feingold. Cambridge: Cambridge University Press, 1990. 1-104.
The Mathematicians Apprenticeship: *Science, Universities and Society in England, 1560-1640.* Cambridge: Cambridge University Press, 1984.
"Newton, Leibniz, and Barrow Too: An Attempt at a Reinterpretation." *Isis* 84 (1993): 310-338.
Fleg, Graham. "Nicholas Chuquet--an Introduction." *Mathematics from Manuscript to Print: 1300-1600*. Ed. Cynthia Hay. Oxford: Clarendon Press, 1988. 59-72.
Fox, Luke. *North-West Fox, or Fox from the North-west passage* . . . London, 1635.
Furley, David."Aristotle and the Atomists on Infinity." *Cosmic Problems: Essays on Greek & Roman Philosophy of Nature.* Cambridge: Cambridge University Press, 1989. 103-114.
Two Studies in the Greek Atomists. Princeton: Princeton University Press, 1967.
Galileo Galilei. *Dialogues Concerning Two New Sciences*. Trans. Henry Crew and Alfonso de Salvio. New York: Dover, 1954.
Gascoigne, John. "Isaac Barrow's Academic Milieu." *Before Newton: The Life and Times of Isaac Barrow.* Ed. Mordechai Feingold. Cambridge: Cambridge University Press, 1990. 250-290.
"A Reappraisal of the Role of the Universities in the Scientific Revolution" in *Reappraisal's of the Scientific Revolution.* ed. David C. Lindberg, Robert S. Westman Cambridge: Cambridge University Press, 1990. 207-260.
Gatti, Hilary. *The Renaissance Drama of Knowledge: Giordano Bruno in England.* London: Routledge, 1989.
Gaukroger, Stephen. *Cartesian Logic: An Essay on Descartes's Conception of Inference.* Oxford: Clarendon Press, 1989.
Gibson, G. A. "Napier's Logarithms and the Change to Briggs's Logarithms." *Napier*

Tercentenary Memorial Volume. Ed. Cargill Gilston Knott. London: Longmans, Green and Co., 1915. 111-137.
Goldstine, Herman H. *A History of Numerical Analysis from the 16th through the 19th Century*. New York: Springer-Verlag, 1977.
Grafton, Anthony. "Barrow as a Scholar," *Before Newton: The Life and Times of Isaac Barrow*. Ed. Mordechai Feingold. Cambridge: Cambridge University Press, 1990. 291-302.
Grendler, Paul F. *Schooling in Renaissance Italy*. Baltimore: John Hopkins University Press, 1989.
Hadden, Richard. W. "The Social Origins of Early Modern Science." Diss. McMaster University, 1984.
Harriot, Thomas. *Artis Analyticae Praxis*. London, 1631.
Bodleian MSS 313 f. 65.
Hawkins, William. "The Mathematical Work of John Napier (1550-1617)." Diss. University of Auckland, 1982.
Heath, Sir. T.L. *Mathematics in Aristotle*. Oxford: Clarendon Press, 1949.
Hedrich Hirsch, David A. "Donne's Atomies and Anatomies: Deconstructed Bodies and the Resurrection of Atomic Theory." *Studies in English Literature: 1500 - 1900*. 31 (1991): 69-94.
Henry, John ."Thomas Harriot and Atomism: A Reappraisal", *History of Science* 20 (1982): 267- 296.
Heilbron, John L "Mechanics." *Dictonary of the History of Science*. 253-256.
Hill, Abraham. "Some Account of the Life of Dr. Isaac Barrow." *Theological Works of Isaac Barrow*. Ed. Alexander Napier. 9 Vol. Cambridge, 1859. 1: xxxvii-liv.
Hill, Christopher. *The Intellectual Origins of the English Revolution*. Oxford: Clarendon Press, 1965.
Hill, Katherine."`Juglers or Schollers?': Negotiating the Role of a Mathematical Practitioner" *British Journal for the History of Science* **31** (1998): 253-74.
"The Rhetoric of Utility: Avoiding Occult Associations for Mathematics through Profitability and Pleasure", *History of Science* **37** (1999):151-178.
"Mathematics as a Tool for Social Change: Educational Reform in Seventeenth-century England", *The Seventeenth Century* **12** (1997):23-36.
Hobson, E. W. *John Napier and the Invention of Logarithms 1614*. Cambridge: Cambridge University Press, 1914.
Howell, Wilber Samuel. *Logic and Rhetoric in England, 1500-1700*. New York: Russell and Russell, 1956.
Howson, Geoffrey. *A History of Mathematics Education in England*. Cambridge: Cambridge University Press, 1982.
Hunter, Michael. *Science and Society in Restoration England*. Cambridge: Cambridge University Press, 1981.
Hofmann, Joseph. *Leibniz in Paris, 1672-1676: His Growth to Mathematical Maturity* Cambridge: Cambridge University Press, 1974.
Huxley, G. "Henry Briggs." *Dictionary of Scientific Biography*. 461-465.
Jacuot, Jean. "Thomas Harriot's Reputation for Impiety." *A Source Book for the Study of Thomas Harriot*. Ed. John Shirley. New York: Arno Press, 1981: 164-187.
"Harriot, Hill, Warner and the New Philosophy." *Thomas Harriot: Renaissance Scientist* Ed. John Shirley. Oxford: Clarendon Press, 1974: 107-128.
Jayawardene, S.A., "The Influence of Practical Arithmetics on the *Algebra* of Rafael Bombelli", *Isis* 64 (1973).
Jesseph, Douglas M. *Squaring the Circle: The War between Hobbes and Wallis*. Chicago: Chicago University Press, 1999.
Johnston, Stephen. "Mathematical Practitioners and Instruments in Elizabethan England." *Annals of Science* 48 (1991): 319-344.
Johnson, F. R. and Larkey, S.V. "Robert Recorde's Mathematical Teaching and the Anti-Aristotelian Movement." *The Huntington Library Bulletin* 7 (1975): 59-87.
Jones, Charles. "The Concept of One as a Number." Diss. University of Toronto, 1978.
Jones, Richard Foster. *Ancients and Moderns: A Study of the Background of the Battle of the Books*. St. Louis: Washington University Press, 1936.
Kahn, David. *The Codebrakers*. New York: Macmillian, 1967.
Kaplan, E."Robert Recorde: Studies in the Life of a Tudor Scientist", PhD dissertation, NewYork University, 1960.

Kargon, Robert. *Atomism in England from Harriot to Newton*. Oxford: Clarendon Press, 1966.
Kearney, Hugh. *Scholars and Gentlemen: Universities and Society in Pre-Industrial Britain: 1500-1700*. London: Faber and Faber, 1970.
Kepler, J. *Gesammelte Werke Band IX Mathematische Scriften*. Ed. Franz Hammer. Munich: C.H. Beck'sche Verlagsbuchhandung, 1960.
Klein, Jacob. *Greek Mathematical Thought and the Origin of Algebra*. Trans. Eva Brann. Cambridge: M.I.T. Press, 1968. The original German text was published as "Die griechische Logistik und die Entstehung der Algebra." *Quellen und Studien zur Greschichte der Mathematik, Astronomie und Physik*, Abteilung B: *Studien* Vol. 3, fasc. 1 (Berlin, 1934), pp. 8-105 (Part I); fasc. 2 (1936), pp. 122-235 (part II).
Kline, Morris. *Mathematical Thought From Ancient to Modern Times*. 3 vols. Oxford: Oxford University Press, 1972.
Knorr, Wilber ""La Croix des Mathématiciens": The Euclidean Theory of Irrational Lines," *Bulletin of the American Mathematical Society* 9 (1983): 41-69.
Lewis, Albert C. "Complex Numbers and Vector Algebra," in *Companion Encyclopedia of the History and Philosophy of the Mathematical Sciences*, ed. I. Grattan-Guiness. New York: Routledge, 1994.
"Harriot on Ballistic Parabolas." *A Source Book for the Study of Thomas Harriot*. Ed. John Shirley. New York: Arno Press, 1981: 230-264.-."Thomas Harriot." *Dictionary of Scientific Biography*. 124-129.
"Thomas Harriot (1560-1621): The Tycho Brahe of Optics" *Centaurus* 6 (1959): 113-121.
McLean, Antonia. *Humanism and the Rise of Science in Tudor England*. New York: Neale Watson Academic Publications, 1972.
Mahoney, Michael. *The Mathematical Career of Pierre De Fermat (1601 - 1665)*. Princeton: Princeton University Press, 1973.
"The Beginnings of Algebraic Thought in the Seventeenth Century." *Descartes Philosophy, Mathematics and Physics*. Ed. Stephen Gaukroger. New Jersey: Barnes and Noble, 1980: 141-155.
"Changing Canons of Mathematical and Physical Intelligibility in the Later 17th Century." *Historia Mathematica* 11 (1984): 417-423.
Mancosu, Paolo. "Descartes's *Géométrie* and Revolutions in Mathematics," in *Revolutions in Mathematics*, ed. Donald Gilles. Oxford: Clarendon Press, 1992.
Masterson, Thomas. *The First Book of Arithmetic*. London, 1592.
McKirahan, Richard Jr. *Principles and Proofs: Aristotle's Theory of Demonstrative Science*. Princeton: Princeton University Press, 1992.
Miller, Fred D. "Aristotle against the Atomists." *Infinity and Continuity in Ancient and Medieval Thought*. Ed. Norman Kretzmann. Ithaca: Cornell University Press, 1982: 87-111.
Minnaert, M.G.J. "Simon Stevin." *Dictionary of Scientific Biography*. 47-51.
Moulton, Lord. "The Invention Of logarithms, Its Genesis and Growth." *Napier Tercentenary Memorial Volume*. Ed. Cargill Gilston Knott. London: Longmans, Green and Co., 1915. 1-32.
Morley, Florian. "Thomas Harriot: 1560-1621." *A Source Book for the Study of Thomas Harriot*. Ed. John Shirley. New York: Arno Press, 1981.
Naux, Charles. *Historie des Logarithmes de Nepier à Euler*. Paris: A. Blanchard, 1966.
Napier, John. *The Construction of the Wonderful Canon of Logarithms*. Trans. William Macdonald. London: Dawsons of Pall Mall, 1966.
A Description of the Admirable Table of Logaritmes. Trans. Edward Wright. London: 1616. Published in facsimile by Da Capo Press. New York: 1969.
Norwood, Mathew. *System of Navigation*. London, 1692.
Nutton, Vivian. "Greek Science in the sixteenth-century Renaissance." *Renaissance and Revolution: Humanist, Scholars, Craftsmen & Natural Philosophers in Early Moder Europe*. Ed. J.V. Field & Frank A.J.L. James. Cambridge: Cambridge University Press, 1993. 15-28.
Ong, Walter J. *Ramus: Method and the Decay of Dialogue*. Cambridge Mass.: Harvard University Press, 1983.
Osmond, Percey H. *Isaac Barrow: His Life and Times*. London: the Society for Promoting Christian Knowledge, 1944.
Oughtred, William."To the English Gentrie. . .the Just Apologie of William Oughtred." *The Circles of Proportion and the Horizontal Instrument. Both invented, and the Uses of both Written in Latine by Mr. W.O.* Trans. William Foster London, 1634.

Clavis Mathematicae. Trans. Robert Wood. London, 1647.
Pedersen, Olaf. "Logistics and the Theory of Functions: An Essay in the History of Greek Mathematics." *Archives Internationales d'Histore des Sciences* 24 (1974): 29-50.
Early Physics and Astronomy: A Historical Introduction. Cambridge: Cambridge University Press, 1993.
Peter Pesic, "Secrets, Symbols, and Systems: Parallels between Cryptanalysis and Algebra, 1580-1700" *Isis* 88 (1997): pp.674-692.
Proclus. *A Commentary on the First Book of Euclid's Elements*. Trans. Glenn Morrow. Princeton: Princeton University Press, 1970.
Pycior, Helena. "Mathematics and Philosophy: Wallis, Hobbes, Barrow, and Berkeley." *Journal of the History of Ideas* 42 (1987): 265-286.
Symbols, Impossible Numbers, and Geometric Entanglements: British Algebra through the Commentaries on Newton's Universal Arithmetick. Cambridge, 1997.
Recorde, Robert. *The Whetstone of Witte*. London, 1557.New York: Da capo Press, 1956.
The Ground of Artes. London, 1543.
The Pathway to Knowledge. London, 1551.
The Castle of Knowledge. London, 1556.
Rigaud, Stephen. *Correspondence of Scientific Men of the Seventeenth Century*. Oxford: Oxford University Press, 1841.
Rose, Peter. *The Italian Renaissance of Mathematics*. Geneve: Librairie Droz, 1975.
Rosen, Edward. "Harriot's Science: The Intellectual Background," in *Thomas Harriot: Renaissance Scientist*. Ed. John Shirley. Oxford: Clarendon Press, 1974: 1-15.
Ross, Sir David. *Aristotle*. London: University Paperbacks, 1964.
Sasaki, Chikara. "The Acceptance of the Theory of Proportions in the Sixteenth and Seventeenth Centuries: Barrow's Reaction to the Analytic Mathematics." *Histora Scientiarum* 29 (1985): 83-116.
Sampson, R. A. "Bibliography of Books Exhibited at the Napier Tercentenary Celebration, July 1914." *Napier Tecentebary Memorial Volume*. Ed. C.G. Knott. London: Longsmans, Green & Co., 1915: 177-242.
Schmitt, Charles B. *Aristotle and the Renaissance*. Cambridge: Harvard University Press, 1983.
Schonbeck, J. G. "Euclidean and Archimedian traditions in the Middle Ages and the Renaissance." *Companion Encyclopedia of the History and Philosophy of the Mathematical Sciences*. London: Routledge, 1994:173-184.
Scott, J. F. "John Wallis as a Historian of Mathematics." *Annals of Science*. 1, No.3. (1936): 335-357.
"The Reverend John Wallis, F.R.S. (1616-1703)." *Notes and Records, Royal Society of London* 15 (1960): 57-67.
Scriba, Christopher. "The Autobiography of John Wallis, F.R.S." *Notes and Records, Royal Society of London* 25 (1970): 17-46.
Studien zur Mathematik des John Wallis (1616 - 1703). Wiesbaden: Franz Steiner, 1966.
"A Tentative Index of the Correspondence of John Wallis, F.R.S." *Notes and Records, Royal Society of London* 22 (1967): 58-93.
Sesiano, Jacques. "On an Algorithm for the approximation of Surds from a Provencal Treatise." *Mathematics from Manuscript to Print: 1300-1600*. Ed. Cynthia Hay. Oxford: Clarendon Press, 1988: 30-58.
Shapin, Steven. *A Social History of Truth: Civility and Science in Seventh-Century England*. Chicago: Univeristy of Chicago Press, 1994.
Shennan, Francis. *Flesh and Bones*. Edinburgh: Napier Polytechnic, 1989.
Shirley, John. "The Scientific Experiments of Sir Walter Ralegh, the Wizard Earl, and the Three Magi in the Tower, 1603-1617." *A Source Book for the Study of Thomas Harriot*. Ed. John Shirley. New York: Arno Press, 1981. *Ambix* 4 (1949): 52-66.
Smith, David. "The law of Exponents in the Works of the Sixteenth Century." *Napier Tercentenary Memorial Volume*. Ed. C.G. Knott. London: Longsmans, Green & Co., 1915: 81-91.
Smith, David E. "John Wallis as a Cryptographer, "*Bulletin of the American Mathematical Society, 24* (1917): pp. 83-96; pp. 166-169.
Sorabji, Richard. "Atoms and Time Atoms." *Infinity and Continuity in Ancient and Medieval*

Thought Ed. Norman Kretzmann. Ithaca: Cornell University Press, 1982: 37-86.
Steggall, J. E. A. "A Short Account of the Treatise *De Arte Logistica*." *Napier Tercentenary Memorial Volume*. Ed. Cargill Gilston Knott. London: Longsmans, Green and Co., 1915: 145-161.
Stevin, Simon. *The Principle Works of Simon Stevin*. 3 Vols. Ed. D.J. Struik. Amsterdam: C.V. Swets and Zeitlinger, 1958.
Stones, G. B. "The Atomic View of Matter in the Xvth, XVIth, and XVIIth Centuries." *Isis* 10, 1 (1928): 445-65.
Struick, D. J. "Kepler as a Mathematician." *Johann Kepler 1571 - 1630 : A Tercentenary Commemoration of his Life and Work.* Baltimore: The William & Wilkens Co., 1931: 39-57.
Swetz, Frank J. *Capitalism and Arithmetic.* La Salle: Opencourt, 1987.
Tanner, R.C.H. "Henry Stevens and the Associates of Thomas Harriot." *Thomas Harriot: Renaissance Scientist.* Ed. John Shirly. Oxford: Clarendon Press, 1974: 91-106.
"The Ordered Regiment of the Minus Sign: Off-beat Mathematics in Harriot's Manuscripts." *Annals of Science* 37 (1980): 127-158.
Tapp, John. *The Path-Way to Knowledge*. London, 1613.
Tartalea, Nicolo. *Euclides Solo introdttore della scientie mathematice: diligentemente reassettato, et alla integrita ridotto per il degno prefessure di tal scientie.* Vinegina, 1543.
Taylor, E.G.R. *The Mathematical Practitioners of Tudor and Stuart England.* Cambridge: Cambridge University Press, 1954.
Thomas, Keith. "Numeracy in Early Modern England." (The Prothero Lecture). *Transactions of the Royal Historical Society*. 37 (1987): 103-32.
Tillotson, John. *The Works of the Learned Isaac Barrow, D.D. Late Master of Trinity College in Cambridge (Being all his English Works)*. London, 1716.
Turner, A. J. "William Oughtred, Richard Delamain and the Horizontal Instrument in Seventeenth Century England." *Annali Dell'Istituto e Museo di Storia della Scienza di Firenze* 6 (1981): 99-125.
"Mathematical Instruments and the Education of Gentlemen." *Annals of Science* 30 (1973): 51-88.
van Egmond, Warren. "The Commercial Revolution and the Beginnings of Western Mathematics in Renaissance Florence, 1300-1500." Diss. Indiana University, 1976.
van Lohne, J.A. "Thomas Harriot als Mathematiker." *Centaurus* 11 (1965):??
Viète, F. *Introduction to the Analytic Art.* Trans. Reverend J. W. Smith *Greek Mathematical Thought and the Origin of Algebra.* Cambridge: M.I.T. Press, 1968: 315-353.
Wallis, John. *A Treatise of Algebra, Both Historical and Practical. Shewing, The Original Progress. and Advancement thereof, from time to time; and by what Steps it hath attained to the Height at which it now it is.* London: 1685.
Opera Mathematica, 3 Vol. Oxford: 1695.
"An Essay of Dr. Wallis, exhibiting his Hypothesis about the Flux and Reflux of the Sea." *Philosophical Transactions* 16 (1666): 262-281.
Ward, John. *The Lives of the Professors of Gresham College.* 1740. New York: Johnson Reprint Corporation, 1967.
Waterhouse, Edward. *A Declaration of the State of Virginia . . .and a Treatise Annexed, Written by that learned Mathematician Mr Henry Briggs, of the Northwest passage to the South Sea through the Continent of Virginia* . . . .London, 1622.
Waters, David. *The Art of Navigation in England in Elizabethan and Early Stuart Times.* London: Hollis and Cauter, 1958.
Webster, Charles. *The Great Instauration: Science, Medicine, and Reform 1626-1660.* New York: Holmes & Meirer Publications, 1975.
Webster, John. *Academiarum Examen; or the Examination of Academies.* London, 1654.
White, Michael. L. *The Continuous and the Discrete: Ancient Physical Theories from a Contemporary Perspective.* Oxford: Clarendon Press, 1992.
Whiteside, D.T. "Mathematical Thought in the Later 17th Century" *AHES.* 1 (1960): 179-388
"Isaac Barrow." *Dictionary of Scientific Biography.* 473-476.
Willmoth, Frances. "'The Genius of all Arts' and the Use of Instruments: Jonas Moore (1617-1679) as a Mathematician, Surveyor, and Astronomer." *Annals of Science* 48 (1991): 355-366..

"Mathematical Sciences and Military Technology: the Ordnance Office in the Reign of Charles II." *Renaissance and Revolution: Humanist, Scholars, Craftsmen & Natural Philosophers in Early Modern Europe*. Ed. J.V. Field & Frank A.J.L. James. Cambridge: Cambridge University Press, 1993. 117-132.

Sir Jonas Moore: Practical Mathematics and Restoration Science. Woodbridge: Boydell Press, 1993.

Wingate, Edmund. *Arithmetique Made Easie*. (London, 1630.

Wood, Anthony à. *Athenae Oxonienses; An Exact History of All the Writers and Bishops Who Have Had Their Education in the University of Oxford, to Which are Added the Fasti, or Annals of the Said University*. 4 vols. London. Reprint, Sources of Science, no. 55. New York, Johnson Reprint, 1967.

Wright, Louis B. , *Religion and Empire: The Alliance between Piety and Commerce in English Expansion 1558-1625*. Chapel Hill: University of North Carolina Press, 1943.

INDEX

INDEX OF NAMES

INDEX OF TOPICS

AUSTRALASIAN STUDIES IN HISTORY AND PHILOSOPHY OF SCIENCE

1. R. McLaughlin (ed.): *What? Where? When? Why?* Essays on Induction, Space and Time, Explanation. Inspired by the Work of Wesley C. Salmon. 1982 ISBN 90-277-1337-5
2. D. Oldroyd and I. Langham (eds.): *The Wider Domain of Evolutionary Thought.* 1983 ISBN 90-277-1477-0
3. R.W. Home (ed.): *Science under Scrutinity.* The Place of History and Philosophy of Science. 1983 ISBN 90-277-1602-1
4. J.A. Schuster and R.R. Yeo (eds.): *The Politics and Rhetoric of Scientific Method.* Historical Studies. 1986 ISBN 90-277-2152-1
5. J. Forge (ed.): *Measurement, Realism and Objectivity.* Essays on Measurement in the Social and Physical Science. 1987 ISBN 90-277-2542-X
6. R. Nola (ed.): *Relativism and Realism in Science.* 1988 ISBN 90-277-2647-7
7. P. Slezak and W.R. Albury (eds.): *Computers, Brains and Minds.* Essays in Cognitive Science. 1989 ISBN 90-277-2759-7
8. H.E. Le Grand (ed.): *Experimental Inquiries.* Historical, Philosophical and Social Studies of Experimentation in Science. 1990 ISBN 0-7923-0790-9
9. R.W. Home and S.G. Kohlstedt (eds.): *International Science and National Scientific Identity.* Australia between Britain and America. 1991 ISBN 0-7923-0938-3
10. S. Gaukroger (ed.): *The Uses of Antiquity.* The Scientific Revolution and the Classical Tradition. 1991 ISBN 0-7923-1130-2
11. P. Griffiths (ed.): *Trees of Life.* Essays in Philosophy of Biology. 1992 ISBN 0-7923-1709-2
12. P.J. Riggs (ed.): *Natural Kinds, Laws of Nature and Scientific Methodology.* 1996 ISBN 0-7923-4225-9
13. G. Freeland and A. Corones (eds.): *1543 and All That.* Image and Word, Change and Continuity in the Proto-Scientific Revolution. 1999 ISBN 0-7923-5913-5
14. H. Sankey (ed.): *Causation and Laws of Nature.* 1999 ISBN 0-7923-5914-3
15. R. Nola and H. Sankey (eds.): *After Popper, Kuhn and Feyerabend.* Recent Issues in Theories of Scientific Method. 2000 ISBN 0-7923-6032-X
16. K. Neal: *From Discrete to Continuous.* The Broadening of Number Concepts in Early Modern England. 2002 ISBN 1-4020-0565-2

KLUWER ACADEMIC PUBLISHERS – DORDRECHT / BOSTON / LONDON

MIX
Papier aus verantwortungsvollen Quellen
Paper from responsible sources
FSC® C105338

FSC
www.fsc.org

If you have any concerns about our products,
you can contact us on
ProductSafety@springernature.com

In case Publisher is established outside the EU,
the EU authorized representative is:
Springer Nature Customer Service Center GmbH
Europaplatz 3, 69115 Heidelberg, Germany

Printed by Libri Plureos GmbH
in Hamburg, Germany